Impact of Climate Change on Water Resources in Agriculture

Impact of Climate Change on Water Resources in Agriculture

Editors

Cornélio Alberto Zolin
Brazilian Agricultural Res. Corporation
Embrapa Agrosilvopastoral
Mato Grosso (MT)
Brazil

Renato de A.R. Rodrigues
Brazilian Agricultural Res. Corporation
Embrapa Soils
Rio de Janeiro (RJ)
Brazil

CRC Press
Taylor & Francis Group
Boca Raton London New York

CRC Press is an imprint of the
Taylor & Francis Group, an **informa** business
A SCIENCE PUBLISHERS BOOK

CRC Press
Taylor & Francis Group
6000 Broken Sound Parkway NW, Suite 300
Boca Raton, FL 33487-2742

Printed on acid-free paper
Version Date: 20150528

International Standard Book Number-13: 978-1-4987-0614-8 (Hardback)

Visit the Taylor & Francis Web site at
http://www.taylorandfrancis.com

and the CRC Press Web site at
http://www.crcpress.com

Preface

We were pleased to have been asked to edit this book on "Impact of Climate Change on Water Resources in Agriculture". Undoubtedly this subject poses one of the biggest challenges for food, energy, fiber and water security worldwide and, as a consequence, for society. In this context, the approach to this crucial subject needs to consider different points of view, from simulation of hydrological cycle to a multinational discussion on polices to cope with the possible impacts of climate change scenarios in the near future.

The impact of climate change is difficult to forecast. In this book, we aim to provide an integrated analysis about climate change, water resources and agriculture – all in one text. Such an ambitious goal has the purpose of providing readers with current and inter-multidisciplinary information regarding climate change and water resources in agriculture, as well as to give a comprehensive perspective about the core points which have a bearing on this subject matter.

Despite all the difficulties we faced to accomplish our goal of presenting an international perspective on this broad topic, we made a special effort to bring together insights from different scientists around the globe working with the multifaceted aspects of climate change and water resources in agriculture.

Beside the fundamental topics that are covered in this book, important insights concerning hydrological cycles, forests, land use change and water, low carbon agriculture, green economy concepts, as well as the current state of the United Nations Framework Convention on Climate Change and Global Research Alliance are also provided. Additionally, it seeks to motivate readers to critically examine the theme of this book.

Finally, we would like to thank the Brazilian Agriculture Corporation (Embrapa) and all contributors of this book, without whom we could not have achieved this task.

Editors
Cornélio Alberto Zolin
Renato de A.R. Rodrigues

Contents

1

The Role of Agriculture on the UNFCCC Negotiation Process

Gustavo Barbosa Mozzer[1] and *Simone Aparecida Vieira*[2]

Scientific Background

The intensity, magnitude and extent of human alteration of the landscape is unprecedented (Lambin et al., 2001). Such changes in land use and land cover are so important that they even affect fundamental aspects of planetary dynamics (Silva et al., 2011), such as biodiversity (Sala et al., 2000), influencing regional microclimate (Chase et al., 2000) resulting in draught or soil salinization or contributing to the generation of large-scale effects (Houghton, 1999).

It is crucial to point out how agriculture has played a role in shaping the terrestrial landscape. The binomial agricultural and environmental degradation is the oldest form of natural environment conversion, and it has promoted more changes in the landscape than any other human activity (Clay, 2004). Some pessimistic analysis may suggest that, over the past 30 years, demand for land has grown about 13 million hectares per year, with much of this demand being met by the deforestation of natural areas (FAO, 2012). Depending on specific conditions, changes in land use may result in increased environmental vulnerability and social economic disturbances (White, 1997; Middleton, 1998; Lambin et al., 2001).

Preterit experiences have shown that conventional farming practices with carefree maintenance of environmental services and conservation of soils also reduce productivity as a consequence of environmental degradation (Tilman et al., 2002). It is estimated that, due to misuse, there are in Brazil 61 million hectares of degraded land that could be recovered and used in food production (Silva et al., 2011).

[1] Brazilian Agricultural Research Corporation (Embrapa) - Secretariat for International Affairs, Edifício Embrapa Sede - Prédio CECAT - 3.° Andar, Parque Estação Biológica - PqEB, Av. W3 Norte, Brasília, DF - Brasil.
Email: gustavo.mozzer@embrapa.br

[2] University of Campinas (Unicamp) - Environmental Studies and Research Center – Nepam, Cidade Universitária "Zeferino Vaz", Barão Geraldo - Campinas, São Paulo, Brasil.
Email: sivieira@unicamp.br

Climate change gives rise to important characteristics for the agricultural sector. The dilemma is the need to increase sector productive capacity, aiming to reduce the emissions of greenhouse gases and hence adapt practices and process to linear and nonlinear (Lopes et al., 2011) possible impacts of climate change (Assad et al., 2008; Zullo Junior et al., 2008) or, from another standpoint, the predictable and unpredictable risks (Giddens, 2009) related to the increase of average global temperature.

Politically, it is clear for the need to prioritize food security and consequently universalize access to food. Such gains in net food production could be achieved by the implementation of rational land use associated with the expansion of agricultural fields, or by the increase in overall levels of production efficiency (Assad et al., 2008). In such a context, the priority role that the agricultural sector could achieve in the UNFCCC context is still unclear. In any case the key challenge of the agricultural sector in the near future would most certainly be related to seeking and investing in adaptive strategies or should the agricultural sector mainly contribute to the overall effort to reduce global GHG emissions, and therefore compromise on the ability to maximize the potential to increase food production.

Considering current technological development and the maintenance of agricultural practices and processes, if on one hand the increase in food production would result in an equivalent increase in GHG emissions, on the other hand the incremental negative effects that are expected as a result of climate change would inevitably promote a reduction in the overall potential of agricultural systems to adequately respond to the current and future demands of food production.

Based on the above ideas, it is very clear and indubitably defensible that new agricultural processes and practices, new crop breeds, new varieties, and new food types are certainly strategies that have to be taken into consideration while tackling the role of agriculture and climate change in a medium to long-term strategy. At the same time, it is also reasonable to assume that adaptation should be the most relevant pillar on the agriculture-climate change agenda.

Despite these facts, science and rationality have been obscured when it comes to the implementation of the political agenda under the UNFCCC. This chapter will discuss how agriculture, initially excluded during the initial phase of implementation of the Kyoto mechanism, due to political interests, has recently emerged into a very polarized debate associated with the urgency of diminishing GHG emissions and the search for new opportunities for the agricultural sector.

Political Arena

The role of agriculture in the United Frame Work Convention on Climate Change (UNFCCC) process has changed since the beginning of the implementation of multilateral talks started in the preparation for the Rio 1992 Earth Summit. In order to understand such changes, it is necessary to keep in perspective the rough path that has been trailed for the implementation of this Convention.[1]

[1] A summarized historical timeline of UNFCCC COPs has been included as the Annex I to facilitate the understanding on how events took place.

It is clear that a big gap distances the academic science from the concrete policy results, which effectively are translated into actions or decisions in the context of the UNFCCC process. Particularly in the agricultural sector, during the early 90's and the first decade of 21th century, governments have been lenient or unable to properly implement a systematic agenda on discussions and debates with the civil society regarding climate change, its negative impacts and a proper medium to long term investment strategy for the agricultural sector.

There are many reasons that could have led to the inability to promote systematic climate change discussions in the agricultural sector. Lack of priority or the sense of urgency commonly found on day to day agendas could explain a very important portion of this inability (Giddens, 2009). Nonetheless, it is necessary to note that, not until very recently, it was well known among the agricultural sector that agriculture had no positive role on the climate change agenda, rather than a significant responsibility for the contribution in GHG emissions.

The reason for the asymmetry regarding the relevance of the agricultural sector in terms of contribution to GHG emissions and its role in the UNFCCC political agenda is related to the strategy adopted during the implementation of the Kyoto Protocol at COP 3 in 1997. At that time, the priority was to develop a multilateral instrument that would enable decisive attitudes in regards to tackling climate change. Kyoto has shown that the creation of such an instrument was, however, extremely complex. This has also been influenced by the fact that nations, especially developed countries, find themselves in different conditions regarding the citizen's interest versus the political will to enforce national actions and international commitments to dedicated agendas.

In this context, during the initial architecture of the Kyoto Protocol, the agricultural sector was strategically left aside,[2] giving room for the industrial sector to figure as the basic building block of the new mechanism. The core of this decision was basically related with the relevance regarding fossil fuel consumption of the industrial, energy and transport sectors compared to others. Additionally, but not less important, the technological feasibility to implement a sound monitoring process of CO_2 emissions and energy consumption was also a decisive factor to corroborate the choice of the industrial sector, including transport and energy, over others.

Monitoring CO_2e emissions in the agricultural sector is definitively not as straightforward as it is in the industrial sector; in fact, this still poses some major challenges for science. Though comparable to the agricultural sector, the forest sector received a very distinctive treatment at the very beginning of the implementation of UNFCCC agenda. More on this topic will be addressed in the section "The role of the forest sector in agriculture negotiation" further ahead in this chapter.

[2] Agriculture is treated under the CDM on a specific sectorial scope. Though very limited, some CDM projects related to enteric fermentation and fertilizer efficiency have been implemented in the agricultural sector. Other projects like waste and manure management, biofuel and energy efficiency have also tackled the agricultural sector.

Recent Developments Under UNFCCC and the Agricultural Sector

Until 2007 when the COP 13 MOP 3 took place in Bali, the key political achievement had been the ratification of the Kyoto Protocol that happened on the margin of COP 10 in Buenos Aires. At that moment, jubilation was the major feeling after almost a decade of uncertainties regarding the future of multilateralism as a meaningful strategy to tackle climate change.

At Bali, negotiations on the next steps of multilateral climate change arena were formally initiated (UNFCCC, 2007). The greatest challenge was to foresee a strategic arrangement that could tackle both the overall lack of ambition from developed countries and the political challenge to cope the United States of America to ratify a legally binding instrument. Also relevant was the necessity to develop and adopt a new kind of strategy capable of leveling the playing field regarding commitment and engagement on the reduction of GHG emissions for both developed and developing countries.

As part of the agreement reached in Bali a specific placement for the agricultural issue had been created in the Ad Hoc Working Group on Long-term Cooperative Action under the Convention (AWG-LCA), under the agenda item Cooperative Sectoral Approaches and Sector-Specific Actions, commonly referred, during UNFCCC negotiations as agenda item 1b4. As the negotiation developed and with the materialization of a possible breakthrough in COP 15 MOP 5 in Copenhagen, a deal in agriculture seemed to be manageable, although there were still some sensitive issues to be addressed. The placement of agriculture in the agenda item 1b4 implied the idea of a focus on mitigation, with pleased Annex I countries and created great discomfort for almost all G77 parties.

Prior and during Copenhagen, discussions regarding agriculture were focused on finding an adequate balance between mitigation, adaptation and food security. The agenda item 1b4 was the framing, and work had to take into consideration its limitations and constringency. Trade was a major issue for many parties, especially for developing countries who were analyzing the possible risk that a treatment regarding agriculture under the UNFCCC could imply new or additional commercial burden or result in restrictive measures for international trade on the agricultural sector. Developed countries refused any mention on trade under the agricultural text, leaving this issue unattended.

Aiming to forge such a deal, especially taking into consideration the need to stabilize the concentration of GHGs in the atmosphere, much more ambition had to be brought to the table. Some may say that this level of ambition and coordination among nations had never been seen before. Under that atmosphere there were great expectations that Copenhagen would finally be the proper time and place to accommodate issues regarding agriculture under the UNFCCC context. Such positive perceptions were based on the following aspects:

a) the relevance of the agricultural sector as a source of GHG emissions for many countries
b) the historical lack of systematic talks on the role that agriculture could play under the UNFCCC; and not less relevant

c) the fact that food and nutritional security remains a key challenge for current and future generations.

However, the reality was far more complex than initially foreseeable. Other plurilateral arrangements started to emerge as alternatives or possible fast tracks to address very specific issues in opposition to the systemic or global treatment that was pursued under the UNFCCC. Under this difficult reality a consensual agreement was far from being feasibly reached without important concessions from all parties.

The relations with other multilateral and plurilateral organizations was very sensitive and, in many cases, decisive to block or facilitate some discussions under the UNFCCC. Some very relevant examples are the World Trade Organization (WTO) and the important Doha round trade talks. Both the International Maritime Organizations (IMO) and the International Civil Aviation Organization (ICAL) are also important examples of plurilateral organizations that have positively and negatively influenced talks on the UNFCCC in different circumstances.

The trade issue became especially relevant during Copenhagen as agricultural talks matured into a balanced text (see BOX 1).

BOX 1. Negotiation text during the final stages of Copenhagen—COP 15 MOP 5 (FCCC/AWGLCA/2009/14).

12. All Parties [, reaffirming the objective, principles and provisions of the Convention and taking into account] [specifically taking into account Article 2, Article 3, paragraph 5, and Article 4, paragraph 1(c), of the Convention, and] their common but differentiated responsibilities and their specific national and regional development priorities, objectives and circumstances, [and that adaptation is of paramount importance to ensure food security,] [shall] [make efforts to enhance mitigation] [in the agriculture sector]:

a) Keep[ing] in mind the need to improve the efficiency and productivity of agricultural production systems [when considering mitigation in agriculture], in a sustainable manner, [taking into account the relationship of agriculture [to] [and] food security [and to adaptation], the linkages between mitigation and adaptation, the interests of small and marginal farmers and [indigenous and] traditional knowledge and practices];

b) [[Including] [by] promoting and cooperating] [promote and cooperate] in the research, development, application and diffusion, including transfer, of technologies, practices and processes [[and methodologies] that could contribute to enhance mitigation and adaptation] in the agriculture sector;

c) [Ensuring that cooperative sectoral approaches in the agriculture sector should not result in the creation of international performance standards for the sector or any other measure that may adversely affect sustainable development and result in barriers to or distortion of, the international trade system of goods and products of the agriculture sector;]

d) [Ensuring that cooperative sectoral approaches in the agriculture sector shall not lead to carbon offsets or approaches that adversely impact forest land.]

13. In this respect, requests the SBSTA, at its thirty-second session, to establish a programme of work [to facilitate] [mitigation in the agriculture sector [, considering the links to adaptation]] [on agriculture] [to enhance [the full, effective and sustained] implementation of article 4.1(c)], and invites Parties to submit their views on the [content [and scope] of the] work programme.

An especially delicate issue regarding the agriculture negotiation during Copenhagen was the lack of movement on the Doha round trade talks on the agricultural sector. This can be clearly evidenced by looking at the unagreed text in Box 1 on paragraph 12(c).

The text construction indicated the clear necessity to explicitly present the framing on which agriculture should be dealt as can be observed in the *chapeau* of paragraph 12 though formal quotation of Article 2, Article 3, paragraph 5, and Article 4, paragraph 1(c), of the Convention and to its principles, in especially the common but differentiated responsibilities (CBDR).

In this context, the scenario of having a weak and unbalanced set of decisions in agriculture seemed very problematic, especially considering the fact that WTO had specific regulations on environmentally friendly produces, the Green Box[3] rules that needed to be carefully considered. Under this scenario, the possibility of allowing for negative trade impacts for developing countries was very evident.

Trade and economic issues were a critical component in the agenda in Copenhagen, moving COP/MOP negotiations into highly commercial and financial discussions. The cost to implement the UNFCCC had become politically and economically very high. In agriculture, the sort of outcome that had to be pursued would necessarily have to do with a decision that would allow for understanding how to measure the impact of good practices in crop systems. It would essentially have also to do with how to compare different crops and different cropping systems. How to account for different climate conditions, for instance, how to compare cropping systems from tropical and temperate regions; from sea level and high lands. It would also involve considerations on taking into account different social circumstances, cultural habits and agricultural practices. In other words, such a decision would have to promote international trade, respecting all differences in agricultural production systems and, at the same time, promoting positive incentives to enhance the level of environmental friendliness of agricultural systems.

Despite the adversities, Brazil never diminished its ambitions on a meaningful outcome from COP15/MOP5 (UNFCCC 2009). On that agenda, a very significant program was prepared to be announced during the high-level sections, by the President Luiz Inácio Lula da Silva himself.

Brazil presented during Copenhagen its plan to reduce emissions in the Amazon Forest by 80% until 2020. Among the activities indicated in this plan, the agricultural sector plays a very significant role, including the implementation of integrated livestock, crops and forest systems, zero tillage, biological nitrogen fixation and increase in silviculture system in Brazil. All these activities are components of the Low Carbon Emission Agriculture Plan (ABC), a sectorial component of the National Policy on Climate Change.

Despite the effort, Copenhagen ended without a decision on agriculture. But it was extremely useful to forge a new paradigm in Brazil national policy on climate change. Before Copenhagen, Brazil was always cautious on dealing with mitigation actions or discussing the need to increase the overall ambition for mitigating GHGs under

---

[3] Domestic support for agriculture that is allowed without limits because it does not distort trade or, at most, causes minimal distortion. They also include environmental protection and regional development programmes.

the UNFCCC. The main view was that developed countries, commonly referred to Annex I countries under the UNFCCC, should take the lead and implement ambitious mitigating efforts.

Nonetheless, developed countries never really took the lead; they never implemented ambitious mitigating policies. The positive side from the role of developed countries was the implementation of a very comprehensive monitoring plan. In addition, the EU should be applauded for the effort to implement the Emissions Trading System[4] (EU ETS) and its support for the CDM.

After Copenhagen, there was the fear that the future of the climate regime under the multilateral system had been compromised, but Brazil ended up with moral authority to strongly engage with Annex I parties on the negotiations and substantively have contributed with proposals that facilitated the accommodation of some of the divergences that compromised the outcome in Copenhagen. Strong results in the reduction on Amazon deforestation and a very comprehensive climate change plan credentialed Brazil to act as one of the few countries capable of largely contributing to raise the bar on global GHG mitigations efforts.

During the preparations to Cancun (COP16/MOP6), Brazil's national policy on climate change started to develop, including extensive talks with the civil society. The Ministry of Agriculture and the Presidency of the Republic were strongly engaged in the process of finalizing the Sectorial Plan on Low Carbon Emissions on the Agriculture Sector Plan-ABC[5] (BRASIL, 2011). A revision on the National Policy on Climate Change had also taken place during the same period. It is necessary to highlight the political importance of this meeting, since it had to be shown that multilateralism was still a meaningful path to address climate change as a global issue. In this context, the political mandate was to work out with partners in finding the minimal understandable outcome. The spirit of goodwill and cooperation was perceivable in the text produced during the Tianjin meeting (see BOX 2), this was the last preparatory meeting before COP16 MOP6 in Cancun.

At Cancun, discussion on 1b4 did not progress much, especially because sectorial approaches also involved bunker fuels, and the EU had just implemented a unilateral tax for the civil society aviation. Maritime trade issues were also very high on the agenda, as the International Maritime Organizations (IMO) had approved, without the support of many member states, a very controversial set of decisions.[6] This decision

[4] Works on the 'cap and trade' principle. A 'cap', or limit, is set on the total amount of certain greenhouse gases that can be emitted by the factories, power plants and other installations in the system. The cap is reduced over time so that total emissions fall. In 2020, emissions from sectors covered by the EU ETS will be 21% lower than in 2005.

[5] Plano de Agricultura de Baixa Emissão e Carbono, in Portuguese.

[6] A significant amount of work on technical and operational measures has been carried out in accordance with the work plan, and at MEPC 59 (July 2009) the committee approved to circulate interim guidelines on technical and operational measures which will reduce the GHG emission from ships.
In 2010, MEPC 61 considered amendments to MARPOL Annex VI as a potential manner for introducing non-mandatory technical and operational measures into IMO regulatory regime. MEPC 62 (July 2011) considered and adopted amendments to MARPOL Annex VI for inclusion of regulations on energy efficiency for ships. These amendments added a new Chapter 4 to Annex VI on regulations on energy efficiency for ships making EEDI mandatory for new ships and SEEMP for all ships.

BOX 2. Negotiation text proposed during the Tianjin meeting prior to COP 16 MOP 6.

Preamble

"Affirming that cooperative sectoral approaches and sector-specific actions in agriculture should not limit the ability of developing country Parties to pursue economic and social development and poverty eradication, and, to that end, that it is essential that cooperative sectoral approaches and sector-specific actions in agriculture are undertaken in a manner that is supportive of an open international economic system."

Operative

"Decides that cooperative sectoral approaches and sector-specific actions in agriculture shall be based on best available science, taking fully into account differences between agricultural systems regarding geographic, economic and social conditions and specific national development priorities and circumstances, in particular of developing country Parties, in accordance with equity and common but differentiated responsibilities and in light of the fact that economic and social development and poverty eradication are the first and overriding priorities of developing country Parties."

implemented rules for burden sharing of GHG emissions on the maritime sector without taking into consideration principles and provisions of the UNFCCC. This initiative was seen by many developing countries as a very dangerous precedent that needed to be carefully analyzed, since it could rapidly spread over to other sectors, compromising the applicability of the Convention, especially its principle of common but differentiated reasonability (CBDR) and the obligation of developed countries to provide financial and technological support.

Although long hours of exhaustive negotiations were spent on the production of a very comprehensive text in agriculture (see BOX 3) at the end, no decision was adopted in Cancun. The unsolved issue on the general framework section for the agenda item 1b4 which included bunkers and agriculture was a major drawback. Inasmuch, trade, and differentiation among developed and developing countries in the provision of the Convention and a balanced treatment between adaptation and mitigation were also issues that were on the table.

At Durban, the mood for a position on agriculture was inconspicuous. Some believed that no decision was possible, mainly due to CBDR and trade issues, but there were also signs that a political decision on agriculture would emerge at the high political level segment, since agriculture was a very sensitive political issue for African countries. These readings were even more disturbing, considering the influence that the World Bank, FAO and other plurilateral initiatives had among those countries. At the end, both views proved to be somehow correct. No decision could be achieved at the technical level, and the political high level decided to move the agriculture issue to the Subsidiary Body for Scientific and Technological Advice (SBSTA). This was a strategically very positive move because it detached agriculture from 1b4 and consequently from both Bunker issues and from the mitigation framing.

BOX 3. Negotiation text during the final stages of Cancun—COP 16 MOP 6 (FCCC/AWGLCA/2010/14).

[The Conference of the Parties,

Reaffirming the objective, principles and provisions of the Convention, in particular its Article 2, Article 3, paragraphs 1 and 5, and Article 4, paragraph 1(c),

Bearing in mind [the need to] [the value of] improve the efficiency and productivity of agricultural production systems in a sustainable manner,

Recognizing the interests of small and marginal farmers, the rights of indigenous peoples and traditional knowledge and practices, in the context of [applicable international obligations and taking into account] national laws and national circumstances,

Recognizing that cooperative sectoral approaches and sector-specific actions in the agriculture sector should take into account the relationship between agriculture and food security, the link between adaptation and mitigation and the need to safeguard that these approaches and actions do not adversely affect food security,

[Affirming that cooperative sectoral approaches and sector-specific actions in the agriculture sector should not constitute a means of arbitrary or unjustifiable discrimination or a disguised restriction on international trade,]

1. [Decides that all Parties, with respect to the agriculture sector and taking into account their common but differentiated responsibilities and their specific national and regional development priorities, objectives and circumstances, should promote and cooperate in the research, development, including transfer, of technologies, practices and processes that control, reduce [or prevent] anthropogenic emissions of greenhouse gases, particularly those that improve the efficiency and productivity of agricultural systems [and management of emissions from livestock] in a sustainable manner and those that could support adaptation to the adverse effects of climate change, thereby contributing to safeguarding food security and livelihoods];
2. [[Affirms] [Further decides] that cooperative sectoral approaches and sector-specific actions in the agriculture sector should not constitute a means of arbitrary or unjustifiable discrimination or a disguised restriction on international trade [, in accordance with Article 3, paragraph 5 of the Convention];]
3. Requests the Subsidiary Body for Scientific and Technological Advice to establish, at its thirty-fourth session, a programme of work on agriculture to enhance the implementation of Article 4, paragraph 1(c), of the Convention, taking into account paragraph 1 above;
4. Invites Parties to submit to the secretariat, by 22 March 2011, their views on the content and scope of the work programme;
5. Requests the secretariat to compile these views into a miscellaneous document for consideration by the Subsidiary Body for Scientific and Technological Advice at its thirty fourth session.]

At SBSTA 36, the old issues come onboard once again (BOX 4). The US and other UMBRELLA[7] countries had radically shifted their position regarding adaptation. At Copenhagen, these countries were radically against having adaptation figuring in an

[7] UMBRELLA Group is a loose coalition of non-EU developed countries, which was formed following the adoption of the Kyoto Protocol. Although there is no formal list, the Group is usually composed of Australia, Canada, Japan, New Zealand, Norway, the Russian Federation, Ukraine and the United States of America.

BOX 4. Negotiation text during SBSTA 36 agenda item 9.

1. The Subsidiary Body for Scientific and Technological Advice (SBSTA) recalling Article 9 of the Convention, and on the basis of the principles and provisions of the Convention, in particular common but differentiated responsibilities, initiated, in accordance with decision 2/CP.17, paragraph 75, an exchange of views on issues relating to agriculture.
2. The SBSTA considered the views submitted by Parties on issues relating to agriculture, and noted that earlier work on agriculture was performed under the Convention.
3. The SBSTA noted that there is a need for an assessment of the current state of scientific knowledge on how to enhance the adaptation of agriculture to climate change impacts, while promoting rural development, and productivity of agricultural systems and food security, particularly in developing countries. This should take into account the diversity of the agricultural systems, and the differences in scale, as well as possible adaptation co-benefits.
4. The SBSTA invited Parties and accredited observers to submit to the secretariat, by 17 September 2012, their views on issues referred to in paragraph 3 above for compilation by the secretariat into a miscellaneous document. It further requested the secretariat to prepare a synthesis of these submissions for consideration by the SBSTA at its thirty-seventh session.
5. The SBSTA requested the secretariat to organize an in-session workshop at its thirty-seventh session on the issues referred to in paragraph 3 above taking into account the submissions referred in paragraph 4 above, and to prepare a report on the meeting to be considered during the session.
6. The SBSTA requested that the actions of the secretariat called for in these conclusions be undertaken subject to the availability of financial resources.

agricultural decision. Now, at SBSTA 36, they were advocating a decision to implement workshops on synergies between mitigation and adaptation. It is necessary to highlight the new context that emerged after Durban. During this meeting, it was decided for the continuation of the Kyoto Protocol in parallel to the negotiation of a new arrangement to be implemented after 2020. This new context was possibly the fundamental driver for a very substantial change in the position of UMBRELLA countries regarding the need to have a decision on a work program[8] in Agriculture. Having 2020 as a tangible timeframe added a new variable to the puzzle. In that the new context, accepting the notion of a work program would, very likely, bind its implementation to the 2020 timeframe. On the other hand, developing countries would most certainly request that some issues like adaptation, food security, resilience and trade be prioritized and therefore treated earlier in the work program agenda.

In Doha, the negotiation on agriculture was again reestablished, but, as in Durban, many developed countries wanted to have it moved to the high level. In that sense, not much effort was being spent on the technical negotiation level. The negotiation text achieved during this COP (see BOX 5) reflects the fact that most of the time countries simply reaffirmed their positions without much will to start a drafting mode

[8] Work Programs, in the language of the Convention, are implemented following very specific procedural rules.

BOX 5. Negotiation text during the final stages of Cancun—COP 16 MOP 6 (FCCC/AWGLCA/2010/14).

Recalling Article 9 of the Convention, and [on the basis of] the principles and provisions of the Convention, [in particular common but differentiated responsibilities],

Reaffirming the objective, principles and provisions of the UN Framework Convention on Climate Change, [in particular common but differentiated responsibilities],

Recognizing the fundamental role of agriculture in food production and food security,

Recognizing that agriculture is central to economic and social development and livelihoods [of developing countries] [of all Parties, particularly developing countries, including LDCs], [taking fully into account economic and social development and poverty eradication are the first and overriding priorities of developing countries],

[Recognizing that agriculture is central to livelihoods of people and to the economic and social development of developing countries,]

Recognizing the need to ensure that food production is not threatened by [the adverse effects of] climate change,

Bearing in mind the need to improve [resilience in] agricultural systems and to improve agricultural productivity in a sustainable manner,

[Recognizing the need to address adaptation and mitigation, as well as the synergies between them, in agriculture.]

Considering the views submitted by Parties1 to the Subsidiary Body for Scientific and Technological Advice on issues relating to agriculture [, and noting that earlier work on agriculture 2 was performed under UN Framework Convention on Climate Change].

Recognizing the interests of small and marginal farmers, indigenous and or traditional knowledge and practices, and the role of women in agriculture,

Further recognising the specific economic and social circumstances [and the social and development needs of developing countries and their] [and] specific regional, national and local contexts, [taking into full account the legitimate priority needs of developing countries for the achievement of sustained economic growth and the eradication of poverty].

Further recognizing the diversity and scale of agricultural systems,

1. Requests the Subsidiary Body for Scientific and Technological Advice to undertake scientific and technological work, [consistent with the Subsidiary Body for Scientific and Technological Advice mandate contained in Article 9 of the UN Framework Convention on Climate Change] [on agriculture and climate change, including, inter alia] [[on impacts of Climate Change on Agriculture] [and agriculture on climate change]] [consistent with the Subsidiary Body for Scientific and Technological Advice mandate contained in Article 9, taking into account the commitments in articles 4.3, 4.4 and 4.5 of the UN Framework Convention on Climate Change;]
2. [Further requests the Subsidiary Body for Scientific and Technological Advice to assess the current state of scientific knowledge on the impacts of climate change on agriculture and food security and on how to enhance the adaptation of agriculture to climate change impacts, while promoting rural development, and productivity of agricultural systems and food security, particularly in developing countries. This should take into account the diversity of the agricultural systems, and the differences in scale, as well as possible adaptation co-benefits;]

BOX 5. contd....

BOX 5. contd.

3. [Further requests the Subsidiary Body for Scientific and Technological Advice to assess the state of scientific knowledge of climate change impacts on agriculture and food security, [and of agriculture on climate change;]]
4. [Further requests the SBSTA to identify innovative, efficient and state-of-the-art technologies and know-how [to cope with climate change impacts] [and impact of agriculture on climate change] for improved agricultural production, as well as measures to reduce post-harvest losses, and advise on the ways and means of promoting the development and transfer of such technologies;]
5. [Further requests the SBSTA to advise on scientific programmes, international cooperation in research and development related to [tackling the impacts of] [adaptation and mitigation responses to] climate change [on [and] agriculture], as well as assess capacity needs to strengthen sub-national, national [, regional and global] scientific programs,] and on ways and means of supporting endogenous capacity-building in developing countries;]
6. [Invites Parties and accredited observers to submit to the secretariat, by 25 March 2013, their views on the matters identified in paragraph 1 above and requested the secretariat compile this information for consideration by the Subsidiary Body for Scientific and Technological Advice at its thirty-eighth session;]
7. Requests the Subsidiary Body for Scientific and Technological Advice to report to the Conference of the Parties at its [nineteenth and twentieth] session on the progress of the work undertaken;]
8. [Requests the secretariat to organize [an] [four] [in-session] [or pre-sessional] [or inter-sessional] workshop[s] at the Subsidiary Body for Scientific and Technological Advice at its [thirty-eighth, thirty-ninth, fortieth and forty-first] meeting[s] on the issues referred to in paragraph xx taking into account submissions referred to in paragraph xx and to prepare a report of the workshop[s] to be considered during the session;]
9. Requests the Subsidiary Body for Scientific and Technological Advice [to ensure developing countries representation and participation] and secretariat to [ensure] [invite] [to ensure developing country representation and to invite] that representatives from relevant accredited observers and experts are invited to the workshop;
10. [Requests the secretariat to review and update the 2008 UNFCCC technical paper on challenges and opportunities for mitigation in the agricultural sector to reflect current scientific knowledge and also to include adaptation co-benefits and trade-offs, to be considered by Parties at the Subsidiary Body for Scientific and Technological Advice at its fortieth session;]
11. Requests that the actions of the secretariat called for in these conclusions be undertaken subject to the availability of financial resources.

negotiation section. One very important issue differed Durban from Doha. At Durban, an adopted decision transferred agriculture into the SBSTA and therefore, during Doha, agriculture did not naturally fit under the high-level segment. This technical detail prevented agricultural discussions to move to the high level, and once again the will to have technical discussions on agriculture prevailed over the intention of having a political solution imposed.

Recently, the US has indicated the willingness to initiate a new plurilateral arrangement focusing on the implementation or dissemination of climate friendly agriculture or resilient technology. This is yet another example of how complex and intricate the multilateral negotiations under the UNFCCC have become. Understanding how all those parallel and plurilateral initiatives communicate to each other, and how they are used politically or economically to influence the negotiations under the UNFCCC is undoubtedly a major challenge.

But despite the lack of confidence the second opportunity to discuss agriculture under the Subsidiary Body for Scientific and Technological Advise the SBSTA 38, was conducted in a much more positive mood. During this session a draft conclusion proposed by the Chair (see BOX 6) could be adopted representing a very important step forward in terms of defining the scope of the dialog that agriculture could achieve under the UNFCCC.

BOX 6. Negotiation text during the final stages of SBSTA 38 (Chinese Proposal).

1. The Subsidiary Body for Scientific and Technological Advice (SBSTA) recalling article 9 of the convention, on the basis of the objective, principles and provisions of the convention initiated, in accordance with decisions 2/CP.17, paragraph 75, an exchange of views on issues relating to agriculture,
2. The SBSTA invited all parties and accredited observers to submit views for the compilation by the secretariat and organize session workshop at its 39th session on the current state of scientific knowledge, on how to enhance the adaptation of agriculture to climate change impacts, while promoting rural development, sustainable development and productivity of agriculture systems and food security in all countries, particularly in developing countries. This should take in to account the diversity of the agricultural systems and the differences in scale as well as possible adaptation co-benefits and to prepare a report on the meeting to be considered during the 40th session.

Despite leaving aside very relevant aspects previously defended by the G77, such as a specific treatment for trade and more clear mention of the principle of Common But Differentiated Responsibilities, this decision set on a very positive position to start a dialog in a technical level regarding how to enhance adaptation of diverse types of agricultural systems to the impact of climate change. The text also includes very clearly the necessity to promote the development of the rural area in association with the increase in food productions, agricultural systems and food security.

To address specific issues from developing countries it is visible that the text focuses its main language into adaptation, leaving no room for the word mitigation. The text also formally accept the notion of differentiation when stating the particular vulnerability of developing countries.

On the side of Annex I, it is perceivable that the notion of efficiency in agricultural systems have been very carefully crafted, in addition to the notion of mitigation though the wording "as well as possible adaptation co-benefits...".

The Role of the Forest Sector in Agricultural Negation

Undoubtedly, the implementation of the Kyoto mechanism served as a very important proxy of the challenge posed on humankind regarding the achievement of the UNFCCC's objectives: "the stabilization of greenhouse gas concentrations in the atmosphere at a level that would prevent dangerous anthropogenic interference with the climate system". But during the negotiations to implement this mechanism there were strong interests to allow for a specific treatment on the forest sector, with the creation of atypical modalities of Clean Development Mechanism (CDM) (UNFCCC, 1997) which resulted in the emission of specific carbon credit[9] unities. The reason for such a treatment has more to do with how negotiations take place rather than with its importance, urgency or relevance. In this specific case, the inclusion of a distinct arrangement for the forest sector, under the CDM, had the objective of taking on board the views and positions of Least Developed Countries (LDCs) and Small Island States. These groups and other forested countries did not find their specific needs attended by the conventional CDM.

Considering the poor results accomplished by the forest sector during over a decade of implementation of CDM activities and even looking beyond the UNCCC frontier under the voluntary market, very modest results have been accomplished by the forest sector in terms of promoting a successful market for tCERs, lCERs or other kinds of voluntary credits. Analyzing the historical data of CDM implementation in Brazil (Miguez, 2008) it could be said that the forest sector really benefited from the scheme created under the Kyoto Protocol. In fact, some analysts may argue that the regulation approved under the CDM were so strict and the nature of temporary and long term credits was so restrictive that the forest sector suffered more from the impacts of misinformation, distrust and draconian contracts than really accounted for any meaningful benefit (Shellard and Mozzer, 2011).

With this perspective assuming if the agricultural sector had, at the very beginning of climate change talks, adopted a similar stand to the forest sector, a strategy to mitigate GHG emissions by developing a specific skim and a particular carbon credit unit, it would be reasonable to imagine that the possible outcome would probably have been very similar to what happened with the forest sector. On the other hand, leaving the CDM restricted to the so called industrial sectors, which encompasses all industrialized activities, including the energy and transportation sectors, allowed the Kyoto mechanism to be fully tested and stressed under market pressures, international interests, and national priorities. Recently, looking at the path trailed by the forest sector was always a natural starting point for discussion on agriculture. With this

[9] Certified Emission Reductions (CERs) were established under the CDM, but specifically for the forest sector Temporary Certified Emission Reductions (CER-t) and Long Term Certified Emission Reductions (CER-l) were established. The nature of the first unit (CER) differ from the nature of the other two since emission reductions implemented under CDM are permanent, while emission reductions implemented under forest activity under CDM (CER-t and CER-l) are temporary.

aspect, the already ongoing work that had been initiated on the REDD[10] and REDD+[11] agenda was studied by agricultural negotiators and experts for lessons learned and experience sharing.

Once again, the agriculture and forest sectors had the opportunity to share the momentum and pave paths under the UNFCCC. Nonetheless, dissimilarities between these two sectors were too strong. In order to illustrate some highly sensitive points, it would be relevant to point out that the forest conservation is an activity mainly implemented by a developing country, while the agricultural sector is relevant across the board for all countries despite it's economical circumstances. Basically, all kind of players under the UNFCCC, including developed and developing countries, LDCs, small islands states, oil countries, in synthesis, all countries, have direct or indirect interest in the agricultural sector.

At the same time the commercial issue is much more sensitive in the agricultural sector than in the forest sector. However, one thing that distinguishes both sectors indubitably is food and nutritional security, where the forest sector plays a coadjutant role, specifically associated to the indigenous people and traditional population.

Conclusions

The results achieved by multilateral negotiation have not been always encouraging over the last decade. Nonetheless, the Kyoto Mechanism has proved to be a versatile and effective tool for promoting cooperation between developed and developing countries aiming to intensify actions and activities, which could maximize the use of capital to reduce emissions of greenhouse gases, promoting technology transfer, and enhancing sub-regional sustainable development.

Developed countries were eager to enforce the notion that mitigation was a great opportunity for the agricultural sector. This is a notion that was initially highlighted by FAO and the World Bank, with the concept of Climate Smart Agriculture.[12] Nonetheless, this idea suffered from lack of scientific support and political convergence. Even the FAO did not take too long to evolve the concept of climate smart agriculture into the context of a climate resilient agriculture.

Developing countries were basically seeking room to promote broader discussions regarding the role of the agricultural sector in the UNFCCC negotiations. It is very likely that most developing countries were seeking some recognition that the agricultural sector is more vulnerable to the effects of climate change and that the

[10] REDD talks; have materialized during Montreal (COP 11/MOP 1). The objective was the implementation of a mechanism under the Convention, which allowed for positive incentives for forest conservation, restoration and management.

[11] It went beyond deforestation and forest degradation, and includes the role of conservation, sustainable management of forests and enhancement of forest carbon stocks.

[12] Climate Smart Agriculture was initially proposed by FAO as an acronym for a set of agricultural practices that would result in the increase of soil fertility, aggregation of organic matter and, consequently, increase systemic resilience. There is no package solution, agricultural practices are defined accordingly to local circumstances by a team of experts, considering, economical social and environmental constraints. The concept of Climate Smart Agriculture has been widely spread in the African continent and many Asian countries.

current level of emissions should raise in the near future due to the necessity to attend current and future demands of food production.

Discussions related to trade and agriculture have strongly influenced the lack of room for flexibility. These were basically unsolved issues that were exported from the world trade organization—WTO and consolidated a point of disagreement among developing and developed countries.

It is certain that there will be no simple solution for the treatment of agriculture under the UNFCCC. The Durban Platform has set the time frame. Countries will most likely seek to maximize the relevance of this sector to global climate change policy in a positive way, but without compromises on global food production, food security and, for some countries, even food sovereignty.

Annex I

A Brief Overview of Decisions (Source UNFCCC[13])

COP 1 (Berlin, 1995) 21 decisions, 1 resolution

Parties agreed that the commitments in the Convention were "inadequate" for meeting the Convention's objective. In a decision known as the Berlin Mandate, they agreed to establish a process to negotiate strengthened commitments for developed countries.

COP 2 (Geneva, 1996) 17 decisions, 1 resolution

The Geneva Ministerial Declaration was noted, but not adopted. A decision on guidelines for the national communications to be prepared by developing countries was adopted. Also discussed—Quantified Emissions Limitation and Reduction Objectives (QELROs) for different Parties and an acceleration of the Berlin Mandate talks so that commitments could be adopted at Kyoto.

COP 3 (Kyoto, 1997) 18 decisions, 1 resolution

The Kyoto Protocol, was adopted by consensus. The Kyoto Protocol includes legally binding emission targets for developed country (Annex I) Parties for the six major greenhouse gases, which are to be reached by the period 2008–2012. Issues for future international consideration include developing rules for emissions trading, and methodological work in relation to forest sinks.

COP 4 (Buenos Aires, 1998) 19 decisions, 2 resolutions

The Buenos Aires Plan of Action, focusing on strengthening the financial mechanism, the development and transfer of technologies and maintaining the momentum in relation to the Kyoto Protocol was adopted.

COP 5 (Bonn, 1999) 22 decisions

A focus on the adoption of the guidelines for the preparation of national communications by Annex I countries, capacity building, transfer of technology and flexible mechanisms.

[13] http://unfccc.int/documentation/decisions/items/2964.php

COP 6 (The Hague, 2000) 4 decisions, 3 resolutions

Part II of the sixth COP (Bonn, 2000) 2 decisions

Consensus was finally reached on the so-called Bonn Agreements. Work was also completed on a number of detailed decisions based on the Bonn Agreements, including capacity-building for developing countries and countries with economies in transition. Decisions on several issues, notably the mechanisms land-use change and forestry (LULUCF) and compliance, remained outstanding.

COP 7 (Marrakesh, 2001) 39 decisions, 2 resolutions

Parties agreed on a package deal, with key features including rules for ensuring compliance with commitments, consideration of LULUCF Principles in reporting of such data and limited banking of units generated by sinks under the Clean Development Mechanism (CDM) (the extent to which carbon dioxide absorbed by carbon sinks can be counted towards the Kyoto targets). The meeting also adopted the Marrakesh Ministerial Declaration as an input into the World Summit on Sustainable Development in Johannesburg.

COP 8 (New Delhi, 2002) 25 decisions, 1 resolution

The Delhi Ministerial Declaration on Climate Change and Sustainable Development reiterated the need to build on the outcomes of the World Summit.

COP 9 (Milan, 2003) 22 decisions, 1 resolution

Adopted decisions focus on the institutions and procedures of the Kyoto Protocol and on the implementation of the UNFCCC. The formal decisions adopted by the Conference intend to strengthen the institutional framework of both the Convention and the Kyoto Protocol. New emission reporting guidelines based on the good-practice guidance provided by the Intergovernmental Panel on Climate Change were adopted to provide a sound and reliable foundation for reporting on changes in carbon concentrations resulting from land-use changes and forestry. These reports were due in 2005. Another major advance was the agreement on the modalities and scope for carbon absorbing forest-management projects in the clean development mechanism (CDM). This agreement completes the package adopted in Marrakesh two years ago and expands the CDM to an additional area of activity. Two funds were further developed, the Special Climate Change Fund and the Least Developed Countries Fund, which will support technology transfer, adaptation projects and other activities.

COP 10 (Buenos Aires, 2004) 18 decisions, 1 resolution

Parties gathered at COP-10 to complete the unfinished business from the Marrakesh Accords and to reassess the building blocks of the process and to discuss the framing of a new dialogue on the future of climate change policy. They addressed and adopted numerous decisions and conclusions on issues relating to: development and transfer of technologies; land use, land use change and forestry; the UNFCCC's financial mechanism; Annex I national communications; capacity building; adaptation and response measures; and UNFCCC Article 6 (education, training and public awareness) examining the issues of adaptation and mitigation, the needs of least developed countries (LDCs), and future strategies to address climate change.

COP 11 (Montreal, 2005) 15 decisions and 1 resolution

COP 11 addressed issues such as capacity building, development and transfer of technologies, the adverse effects of climate change on developing and least developed countries, and several financial and budget-related issues, including guidelines to the Global Environment Facility (GEF), which serves as the Convention's financial mechanism. The COP also agreed on a process for considering future action beyond 2012 under the UNFCCC.

COP 12 (Nairobi, 2006) 9 decisions and 1 resolution

A wide range of decisions were adopted at COP 12 designed to mitigate climate change and help countries adapt to its effects. There was agreement on the activities for the next few years under the "Nairobi work programme on Impacts, Vulnerability and Adaptation", as well as on the management of the Adaptation Fund under the Kyoto Protocol. Parties welcomed the "Nairobi Framework", which will provide additional support to developing countries to successfully develop projects for the CDM. Parties in Nairobi also adopted rules of procedure for the Kyoto Protocol's Compliance Committee, making it fully operational.

COP 13 (Bali, 2007) 14 decisions and 1 resolution

COP 13 adopted the Bali Road Map as a two-year process towards a strengthened international climate change agreement. The Bali Road Map includes the Bali Action Plan that was adopted by Decision 1/CP.13. It also includes the Ad Hoc Working Group on Further Commitments for Annex I Parties under the Kyoto Protocol (AWG-KP) negotiations and their 2009 deadline, the launch of the Adaptation Fund, the scope and content of the Article 9 review of the Kyoto Protocol, as well as decisions on technology transfer and on reducing emissions from deforestation.

COP 14 (Poznan, 2008) 9 decisions and 1 resolution

COP 14 launched the Adaptation Fund under the Kyoto Protocol, to be filled by a 2% levy on projects under the Clean Development Mechanism. Parties agreed that the Adaptation Fund Board should have legal capacity to grant direct access to developing countries. Further progress was made on a number of issues of particular importance to developing countries, including adaptation, finance, technology, REDD and disaster management. COP 14 also saw Parties endorse an intensified negotiating schedule for 2009.

COP 15 (Copenhagen, 2009) 13 decisions and 1 resolution

The Copenhagen Climate Change Conference raised climate change policy to the highest political level, with close to 115 world leaders attending the high-level segment. It produced the Copenhagen Accord, which was supported by a majority of countries. This included agreement on the long-term goal of limiting the maximum global average temperature increase to no more than 2 degrees Celsius about pre-industrial levels, subject to a review in 2015. A number of developing countries agreed to communicate their efforts to limit greenhouse gas emissions every two years. On long-term finance, developed countries agreed to support a goal of mobilizing US$100 billion a year by 2020 to address the needs of developing countries.

COP 16 (Cancun, 2010) 12 decisions and 1 resolution

COP 16 produced the Cancun Agreements. Among the highlights, Parties agreed to: commit to a maximum temperature rise of 2 degrees Celsius above pre-industrial levels; make fully operational by 2012 a technology mechanism to boost the development and spread of new climate-friendly technologies; establish a Green Climate Fund to provide financing for action in developing countries via thematic funding windows. They also agreed on a new Cancun Adaptation Framework, which included setting up an Adaptation Committee to promote strong, cohesive action on adaptation.

COP 17 (Durban, 2011) 19 decisions and 1 resolution

At COP 17, Parties decided to adopt a universal climate agreement by 2015, with work beginning under a new group called the Ad Hoc working Group on the Durban Platform for Enhanced Action (ADP). Parties also agreed a second commitment period of the Kyoto Protocol from 1 January 2013. A significantly advanced framework for the reporting of emission reductions for both developed and developing countries was also agreed, taking into consideration the principle of common but differentiated responsibilities.

COP 18 (Doha, 2012) 26 decisions and 1 resolution

AT COP 18, governments set out a timetable to adopt a universal climate agreement by 2015, to come into effect in 2020. They completed the work under the Bali Action Plan to concentrate on new work towards a 2015 agreement under a single negotiating stream, the ADP. Governments emphasized the need to increase their ambition to cut greenhouse gases and to help vulnerable countries to adapt. COP 18 also saw the launch of a second commitment period under the Kyoto Protocol, from 1 January 2013 to 31 December 2020, with the adoption of the Doha Amendment to the Kyoto Protocol.

Acknowledgments

I would like to thank Mariana Cariello and Giampaolo Queiroz Pellegrino for reviewing this chapter.

References

Assad, E.D. Pinto, J.R. Hilton Silveira Zullo, F.R. Jurandir Marin and G.Q. Pellegrino. 2008. Mudanças Climáticas: impactos físicos e econômicos. Plenarium, v. 5: 96–117.

Brasil. 2011. Plano de Agricultura de Baixa Emissão de Carbono (Plano ABC).

Chase, T.N. et al. 2000. Simulated impacts of historical land cover changes on global climate in northern winter. Climate Dynamics 16(2-3): 93–105, ISSN 0930-7575. Available at: < <Go to ISI>://000085374600002 http://www.springerlink.com/content/g19e2rmwyktr08eu/fulltext.pdf >.

Clay, J. 2004. World Agriculture and the Environment: A Commodity-By-Commodity Guide to Impacts and Practices. Island Press, Washington, DC.

[FAO]. Food and Agriculture Organization of the United Nations. 2012. FAOSTAT Statistics Database. ORGANIZATION, U. F. A. A., Rome.

Giddens, A. 2009. The Politics of Climate Change. Polity Press, Cambridge.

Houghton, R.A., Hackler, J.L. and Lawrence, K.T. 1999. The U.S. carbon budget: contribution from land-use change. Science 285: 574–578, ISSN 285.5427.574.

Lambin, Eric F., Turner, B. L., Geist, Helmut J., Agbola, Samuel B., Angelsen, Arild, Bruce, John W., Coomes, Oliver T., Dirzo, Rodolfo, Fischer, Günther, Folke, Carl, George, P. S., Homewood, Katherine, Imbernon, Jacques, Leemans, Rik, Li, Xiubin, Moran, Emilio F., Mortimore, Michael, Ramakrishnan, P. S., Richards, John F., Skånes, Helle, Steffen, Will, Stone, Glenn D., Svedin, Uno, Veldkamp, Tom A., Vogel, Coleen, Xu, Jianchu. 2001. The causes of land-use and land-cover change: moving beyond the myths. Global Environmental Change, v. 11, 4: 261–269, ISSN 0959-3780. Available at: < http://www.sciencedirect.com/science/article/pii/S0959378001000073 http://ac.els-cdn.com/S0959378001000073/1-s2.0-S0959378001000073-main.pdf?_tid=b2b4985f32938318d6238b3cd585754e&acdnat=1338916931_8f675674b1568534bb2e15c602dc6d8b>.

Lopes, C.A. et al. 2011. Uma análise do efeito do aquecimento global na produção de batata no Brasil. Horticultura Brasileira 29(1): 7–15, ISSN 0102-0536.

Middleton, N. 1998. Regions at risk: comparisons of threatened environments. Progress in Human Geography 22(1): 145–146, ISSN 0309-1325. Available at: < <Go to ISI>://000072994200027>.

Miguez, J.D.G., Oliveira Filho, H.M., Mozzer, G.B. and Magalhães, D.A. 2008. Mitigation actions in Brazil. Plenarium 5: 128–139.

Sala, O.E. et al. 2000. Biodiversity—Global biodiversity scenarios for the year 2100. Science 287(5459): 1770–1774, ISSN 0036-8075. Available at: < <Go to ISI>://000085775300030 http://www.sciencemag.org/content/287/5459/1770.full.pdf >.

Silva, J.A.A., Nobre, A.D., Manzatto, C.V., Joly, C.A., Rodrigues, R.R., Skorupa, L.A., Nobre, C.A., Ahrens, S., May, P.H., Sá, T.D.A., Cunha, M.C. and Rech Filho, E.L. 2011. O Código Florestal e a Ciência: contribuições para o diálogo. Sociedade Brasileira para o Progresso da Ciência (SBPC). Academia Brasilseira de Ciências, São Paulo.

Shellard, S.N. and Mozzer, G.B. 2011. Opportunities and challenges for the implementation of REDD mechanisms. pp. 179–193. *In*: Climate Change in Brazil. IPEA, Brasília.

Tilman, D. et al. 2002. Agricultural sustainability and intensive production practices. Nature 418(6898): 671–677, ISSN 0028-0836. Available at: < <Go to ISI>://000177305600053 http://www.nature.com/nature/journal/v418/n6898/pdf/nature01014.pdf >.

[UNFCCC]. United Nations Framework Convention on Climate Change. 1997. Protocolo de Quioto. Available at: <http://unfccc.int/resource/docs/cop3/07a01.pdf>.

[UNFCCC]. United Nations Framework Convention on Climate Change. 2007. Plano de Ação de Bali. Available at: <http://unfccc.int/documentation/documents/advanced_search/items/6911.php?priref=600004671>.

[UNFCCC]. United Nations Framework Convention on Climate Change. 2009. Acordo de Copenhague. Available at: <http://unfccc.int/resource/docs/2009/cop15/eng/l07.pdf>.

White, R. 1997. Regions at risk: Comparisons of threatened environments. *In:* Kasperson, J.X., Kasperson, R.E. and Turner, B.L. (eds.). Annals of the Association of American Geographers 87(2): 388–389, ISSN 0004-5608. Available at: < <Go to ISI>://A1997XE27300019 >.

Zullo Junior, J., Assad, E.D. and Pinto, H.S. 2008. Mudanças climáticas e suas consequências no Brasil. Scientific American Brasil 6: 71–77.

2

Policies and Initiatives Related to Water and Climate Change in Agriculture

Case Studies from Brazil and Africa

Juliana Gil[1,*] and *Josey Kamanda*[2]

World Water Resources

Only 2.5% of all water on the planet corresponds to surface freshwater (Gleick, 1993). Of the fraction that is extracted for human purposes, about 70% is used for agriculture, 20% by industry (including power generation) and 10% is used for direct human consumption (UNEP, 2011).

Contrary to what was thought until some decades ago, water resources are not infinite and utilizing them wisely is a condition for their preservation over time—especially in the face of population growth, urbanization, increasing competition among productive sectors, and added pressure from shifts in rainfall patterns as a result of human-induced climate change. The UN World Water Development Report (2003) predicts that, in the next 20 years, water availability will decrease by 30%.

[1] Researcher at the Food Security Centre and PhD candidate at the Institute of Land Use Economics in the Tropics and Subtropics at the University of Hohenheim, Wollgrasweg 43, room 234 - 70593 Stuttgart. Email: juliana_gil@uni-hohenheim.de

[2] Researcher at the Food Security Centre and PhD candidate at the Institute of Social and Institutional Change in Agricultural Development at the University of Hohenheim, Wollgrasweg 43, room 126 - 70593 Stuttgart.
Email: J.Kamanda@uni-hohenheim.de

* Corresponding author

The fact that water is very unequally distributed around the world also helps to explain why approximately one billion people still lack access to safe water. Whereas Brazil and Russia together concentrate more than 12,500 km^3/year, countries like Malta, Qatar and the United Arab Emirates have less than 0.2 km^3/year—which conferred these two groups the designation 'water-rich' and 'water-poor' countries, respectively (FAO, 2003).

Water scarcity is not always exclusively determined by geographic reasons. As advocated by the International Water Management Institute (Molden, 2007) and already agreed upon by several other entities, the definition of scarcity relates water availability to water demand. In this sense, an important distinction has to be made between 'physical scarcity' and 'economic scarcity': countries in the first situation have more than 75% of its river flows allocated to agriculture, industries or domestic purposes (accounting for recycling of return flows), whereas in countries facing the second situation water resources are abundant relative to water use, with less than 25% of water from rivers withdrawn for human purposes and with the presence of malnutrition. This implies that dry areas are not necessarily water-scarce (e.g., Mauritania) and that economic scarcity could benefit from additional blue and green water, but human and financial capacity are limiting.

Part of a region's scarcity can be compensated through the so-called 'virtual water trade'. As defined by Allan (1998), water consumed in the production process of an agricultural or industrial product is the 'virtual water' contained in it. Thus, water-rich and water-poor countries can mutually benefit from the exchange of water-intensive goods—either by profiting from a comparative advantage (in the case of water-rich countries) or relieving the pressure on their own water resources (in the case of water-poor countries). According to the Water Footprint Network (2013), the global volume of international virtual water flows in relation to trade in agricultural and industrial products averaged 2320 billion m^3 per year during the period 1996–2005 (Mekonnen and Hoekstra, 2011). The biggest net exporters of virtual water are found in North and South America (the US, Canada, Brazil and Argentina), southern Asia (India, Pakistan, Indonesia, Thailand) and Australia. The biggest net virtual water importers are North Africa and the Middle East, Mexico, Europe, Japan and South Korea.

It should be emphasized, however, that water virtual trade is not always driven by water relative abundance, but instead by several other factors influencing international trade. For example, some northern European countries are virtual water importers even though they do not face severe water scarcity. As pointed out by Hoekstra and Chapagain (2008), when designing effective water management instruments, policy makers should take into account "the critical link between water management and international trade, considering how local water depletion and pollution are often closely tied to the structure of the global economy". The same authors also question "whether trade can enhance global water use efficiency, or whether it simply shifts the environmental burden to a distant location".

Water withdrawals have tripled over the last 50 years, which is largely explained by the rapid increase in irrigation development stimulated by food demand in the 1970s and by the continued growth of agriculture-based economies (World Bank, 2007). While agriculture is the most water-consuming economic activity, irrigation in particular responds for most of the water withdrawn for agriculture: 278.8 million hectares are equipped for irrigation at the global scale, which corresponds to about 20% of cultivated land and 40% of global food production (WWDR, 2012). Approximately 68% of the total irrigated area is located in Asia, 17 in America, 9 in Europe, 5 in Africa and 1% in Oceania, and smaller irrigation areas are spread across almost all populated parts of the world (FAO Aquastat, 2013).

According to the United Nations (WWDR, 2012), agricultural water withdrawal accounts for 44% of total water withdrawal in OECD countries, but for more than 60% within the eight OECD countries that rely heavily on irrigated agriculture. In the so-called 'BRIC countries' (i.e., Brazil, Russian Federation, India and China) agriculture accounts for 74% of water withdrawals in average; this ranges from 20% in the Russian Federation to 87% in India. In the least developed countries (LDCs), it is more than 90%.

Irrigation efficiency is highly variable among different regions of the world and associated water losses can be quite substantial in some regions. Besides, climate change and pollution from pesticides, soil sediments, excess nutrients (i.e., inorganic fertilizers and livestock manure) and other contaminants (e.g., veterinary products) often compromise water quality.

A recent study conducted by the Directorate for Trade and Agriculture of the Organization for Economic Cooperation and Development states that "the overall economic, environmental and social costs of water pollution caused by agriculture across OECD countries are likely to exceed billions of dollars annually" (OECD, 2013). Even though the accuracy of the calculation methods employed in such estimates may be questioned, there is no doubt that negative externalities result from agricultural water use and might constraint the increase of food production in the future.

Climate change, on the other hand, could affect water supply and agriculture through changes in the seasonal timing of rainfall and snow pack melt, as well as higher incidence and severity of floods and droughts (OECD, 2013). Besides direct effects such as changing rainfall patterns, some indirect effects of global warming on vegetation and land cover might equally influence the quality and availability of water in some regions, as will be explained in more detail later.

Water, Climate Change and Agriculture

There is an emerging consensus among scientists that climate warming will intensify, accelerate or enhance the global hydrologic cycle (WHO, UNICEF, 2008) and also have serious impacts on the availability of freshwater as well as on the distribution of water resources around the world. Together with several other climate-related effects, this will certainly affect plant growth and agriculture as a whole—particularly irrigated agriculture, inland fisheries and aquaculture, livestock grazing and fodder production, among other farming systems that depend on water management.

Although measurement techniques for assessing the impacts of climate change on agriculture remain relatively vague, there is no doubt that such impacts cannot be neglected and might produce serious consequences given the intrinsic relationship between water, climate change and agriculture. Besides, a feedback loop would be created, in the sense that these impacts would be mutually reinforcing: on the one hand, unsustainable agricultural practices aggravate climate change by emitting greenhouse gases through processes like enteric fermentation of ruminant herbivores (CH_4), manure generation (CH_4 and N_2O), conventional tillage practices (CO_2), wet rice cultivation (CH_4), burning of agricultural residues (CO_2, CH_4, N_2O, among others), nitrous oxide emissions from the soil due to incorrect application of some fertilizers, consumption of fossil fuels in field operations and transportation of inputs/outputs (CO_2), etc.; on the other hand, changing climatic conditions will certainly affect the quality, distribution and availability of natural resources (including water) necessary for cultivating crops and livestock—eventually encouraging weed and pest proliferation (Cordeiro et al., 2011).

As pointed out by the report *Climate Change—Impacts on Agriculture and Costs of Adaptation* (IFPRI, 2009), "although there will be gains in some crops in some regions of the world, the overall impacts of climate change on agriculture are expected to be negative, threatening global food security". It also highlights that yield declines are expected to occur in developing countries for some of the most important crops—South Asia being the region to be particularly hard hit—as well as the effect that this may have in pushing prices of rice, wheat, maize, soybean and meat upwards.

It is also worth highlighting that increasing animal protein consumption rates in developing countries—especially China and Brazil—will require more water, given that meat production is one of the most water-intensive agricultural activities. According to the FAO report *Livestock Long Shadow*, "the livestock sector is a key player in increasing water use, accounting for over 8 percent of global human water use, mostly for the irrigation of feed crops. It is probably the largest sectorial source of water pollution, contributing to eutrophication, 'dead' zones in coastal areas, degradation of coral reefs, human health problems, emergence of antibiotic resistance and many others. The major sources of pollution are from animal wastes, antibiotics and hormones, chemicals from tanneries, fertilizers and pesticides used for feed crops, and sediments from eroded pastures. (…) Livestock also affect the replenishment of freshwater by compacting soil, reducing infiltration, degrading the banks of watercourses, drying up floodplains and lowering water tables. Livestock's contribution to deforestation also increases runoff and reduces dry season flows" (FAO, LEAD, 2006).

Developing countries are the least resilient to varying climate conditions and are generally found in a much more vulnerable position than developed countries where additional irrigation can be provided during drier periods and risks can be reduced through crop insurance schemes. The fact that developing economies heavily rely on agriculture aggrevates this situation. Sophisticated adaptations have been developed to allow cropping in arid and semi-arid conditions (e.g., mixed and companion cropping, floodwater spreading and runoff harvesting) but most farmers in developing countries do not have access to such technologies. Conservation techniques that may enhance storage of rainwater in the plants root zone—and even enhance carbon uptake by agriculture—are comparatively more (…) accessible. Examples include zero tillage (…) (which is already a common practice in Brazil); mulching (using dust, crop residues or plastic sheeting); occasional deep tillage to deepen the effective root zone and increase its porosity; strip planting; surface shaping to enhance retention and infiltration of rainfall and runoff; and the addition of soil amendments that improve structure and structural stability of the root zone (Turral, H., Burke, J. and Faurès, J. 2011, 23 pp.). Still, these techniques need to be better disseminated among farmers.

Policy Issues Related to Water

As stated by Turral et al., "allocation systems have to smooth out short-term variability in supply and meet longer-term development objectives" (Turral et al., 2012). Aligned with this idea, an increasing number of countries have been trying to implement technical and political strategies aimed at ensuring a rational and fair appropriation of water resources by all stakeholders.

From the technical point of view, the most common measures that countries may take involve financial support for biotechnology development, agricultural innovations and enhancement of water productivity; from a policy perspective, the adoption of economic-oriented instruments, command-and-control policies (regulations), institutional reform, capacity building and investments in infrastructure merit attention.

Although the establishment of economic incentives often meets political opposition (making it hard for some national governments to implement them), they were widely adopted over the last decades because they require minor investments in infrastructure while at the same time produce results almost immediately after coming into force. In addition, economic instruments can be applied individually or coupled with instruments of another nature, and can be an interesting approach when transboundary issues are involved.

Taxes and quotas (or 'permits') are two examples of economic incentives, which differ in terms of efficiency and distributional implications. Volumetric water taxes, for instance, may work well when water flows are not recoverable; otherwise, it does not take into account crop-specific differences in field irrigation efficiency (e.g., taxation on rice is too heavy due to the requirement of flooding paddy fields) and may end up acting as a perverse incentive. Moreover, it is often the case that tax-based policies lead to abusive water utilization, like in the case that wealthier stakeholders may take up almost all available water (after all, they have the means to pay for it) and not enough is left for poorer households to fulfill their needs. The positive relationship between tax payments and water use (i.e., yields tend to increase as water application increases, except in the case of over-irrigation) does not necessarily make the farmer choose the socially optimal water-yield production technique.

Quota-based policies are an alternative to the application of taxes for water-use regulation. Among other advantages, quotas are usually more effective in ensuring water protection because they are limited according to the volume of water that can be withdrawn from water bodies without threatening their integrity (whereas polluters could choose to pay more taxes and continue polluting the environment if that were cheaper in the end). In certain quota systems, permits can be traded among stakeholders in a self-regulating mechanism. However, determining the 'socially optimal distribution' can be quite difficult in practice and costs related to quotas' initial assignment, administration and monitoring may be very high (especially because enforcement control requires measurements on an individual basis). Besides, a quota-based system by itself would not generate the revenue to cover operational and maintenance costs related to water treatment. These considerations lead to the conclusion that, in specific setups (e.g., smaller agricultural communities where farmers use water for irrigation) such systems would be more likely to work, particularly if farmers trust each other and pursue the enforcement of regulations themselves; otherwise, it would probably be difficult to make them agree on the criteria to be used for the initial assignment of quotas, and to ensure transparency throughout the process.

Considering the time gap between scientific knowledge generation and its incorporation into national environmental legislations, positive market incentives have become quite popular in many countries for water management. Instead of establishing obligations and punishing those who do not comply with them, such as tax-based systems, these mechanisms offer a compensation for adopters of best water management practices, such as tax exemptions or remuneration through payment for environmental services schemes (PSA schemes). For instance, indirect subsidies can be an interesting strategy for situations where different sectors/activities compete for water resources, or when the government wishes to reallocate agriculture to more productive areas, stimulate crop diversification, promote the dissemination of less water-intensive varieties or increase the utilization of more efficient irrigation equipment. According to Turral et al. (2011, 114 pp.) "At national level, there are a number of options to adjust the focus and balance of agricultural water management. Investment and subsidies can follow shifts in agro-ecological zones, and can focus on areas that continue to have comparative advantage". This might make sense in terms of food security, but is less likely to deal with problems of social equity.

After analyzing all these options, it is clear that there is no blueprint for regulation on water use and solutions should be designed for each and every context. The identification of the most suitable solution requires a thorough understanding of whether a trade-off is implied concerning efficacy, efficiency and equity aspects of each water management scheme. Of course it is not possible to foresee all outcomes of a policy given that it usually depends on unpredictable stakeholders' behavioral changes and oscillating macro-economic conditions; nevertheless, it is crucial to consider the economic and social impacts it may have prior to its design and implementation.

One way of capturing the differences in the value of water in alternative water uses is to measure the so-called 'water productivity' associated to each of them; this concept refers to amount of a given output (e.g., agricultural production) per input unit of water and the value associated to such output. As Gersfelt (2007) points out, economic efficiency of water usage may be improved by favoring high-value crops and adjusting production techniques (e.g., water-stress) whilst social economic efficiency requires the evaluation of social prices rather than market prices (since the latter may not reflect the true social costs and benefits).

Brazil and Ethiopia: Case Studies

When the climate change–water–agriculture issue is analyzed, Latin America and Africa draw attention for the fact that most of their countries' economies heavily rely on agriculture in terms of income and job generation. Besides, food production is expected to expand in both continents as a result of increasing demand for food in other parts of the world, combined with local availability of unexploited lands (which is a scarce resource elsewhere).

Within Latin America, Brazil deserves to be highlighted as a major agricultural producer, where 12% of the freshwater available worldwide are concentrated but a number of households still face water scarcity. Some studies recently developed by Cepagri/Unicamp (Centre for Meteorological and Climate Research Applied to Agriculture) and Embrapa Agricultural Informatics have tried to simulate the impacts of climate change on agriculture by applying mathematical modeling techniques. Results indicate that the economic losses provoked by a 1 degree Celsius-rise in the average temperature would reach US$ 375 million per year in the case of coffee (considering only what is produced in the states of Minas Gerais, Paraná and São Paulo) and US$ 61 million per year in the case of corn (considering what is produced in São Paulo state) (Assad, 2007).

A more recent study conducted by the same authors (Assad and Pinto, 2008) corroborates earlier predictions and suggests that losses caused by temperature rises may reach BRL $ 7.4 billion in Brazil by 2020. The geography of agricultural production will be altered as a whole, including the migration of species to regions where they do not currently occur (like cassava and coffee, which may disappear from the semi-arid in Southeast Brazil, respectively). Federal states in the South might face water stress, leading to yield reduction for soya and other species adapted to the local tropical weather. Sugar cane, on the other hand, may spread across new areas to the point that its cultivation area will double. Although models and scenarios for different temperature and precipitation conditions are still to be improved, it is already known that agricultural production will suffer.

The inefficiency that generally characterizes irrigation in Brazil may aggravate this situation: it is estimated that only 40% of the water withdrawn by agriculture is effectively utilized, whilst the rest is lost because of bad management and lack of technical expertise. Irrigation methods applied in 93% of Brazilian agriculture are the least efficient ones, such as furrow and sprinkler irrigation; more efficient methods (such as drip irrigation systems) are used in only 4% of the irrigated lands (Ricardo and Campalini, 2007).

On the other hand, "Africa is one of the most vulnerable continents to climate change and climate variability, a situation aggravated by the interaction of 'multiple stresses', occurring at various levels, and low adaptive capacity" (IPCC, 2007). Among the Least Developed Countries in the continent, Ethiopia might suffer serious impacts given that agriculture responds for a large share of its Gross Domestic Product (GDP), labor force occupation and total exports; it is predominantly run by small-scale farmers who employ large rain-fed cultivation practices highly dependent on climate regulation; and finally the natural characteristics of the region (i.e., extreme hydrological variability and the occurrence of the monsoons) is likely to lead to completely different reactions to climate change even among geographically close regions, making it even more complex and costly to design mitigation and adaptation measures.

As an attempt to overcome such challenges and contribute to the consolidation of a new scenario in water management, Brazil and Africa have been promoting initiatives aimed at improving water use and management. Two illustrative case studies are described in this chapter, as well as successful and unsuccessful experiences related to each of them.

Brazil: The Water Producer Program

With the establishment of the National Water Code in 1934 (i.e., the first compendium of laws that attempted to regulate water use and management in the country), the Brazilian Government brought up the need to use water for industrial purposes and for the generation of hydroelectricity. In 1979, as a result of agricultural expansion, demand for irrigation water grew quite substantially and conflicts started emerging. The National Policy for Irrigation was then established, but the dispute between multiple water uses (i.e., navigation, irrigation, power generation and fisheries and supply) continued triggering conflicts in regions with high water demand. In 1997, a Decree creating the 'Waters Act' and the 'National Water Resources Policy' was approved and incorporated the principle that human supply and animal watering were a priority in situations of conflict and scarcity, as it had been already established a few years earlier by the Brazilian National Constitution in 1988.

The current 'Waters Act' defines the watershed as the administrative unit of the water system, on which the water management system is based in Brazil. Any decision related to water utilization, investments in infrastructure and application of financial resources in water quality is made separately for each of the 12 'hydrographic regions' that together comprehend 200 thousand micro-basins.

The National Water Agency (ANA) is an integral part of the complex institutional structure for water management in Brazil. It is an independent regulatory agency with administrative and financial autonomy within the Ministry of Environment. The bill creating the Agency was approved in June 2000 and converted into Law n. 9.984 in the same year. Since then, the agency has incorporated additional attributions, including regulation of water use in federal water bodies, implementation of the National Policy of Water Resources (whose goal is to ensure the democratic and decentralized management of water resources), regulation of irrigation services under concession (Law n. 12.058/2009) and dam safety monitoring.

With the support of other governmental and non-governmental partners, ANA co-developed the so-called 'Water Producer Program', aimed at disseminating payment for environmental services (PAS) (…) schemes in water basins of strategic importance for water conservation (…). Through this program, projects which are able to reduce erosion and siltation of water sources in rural areas—thus contributing to enhance water quality and regulate water supply in the watershed—may receive technical and financial support, as well as a certificate at the end of an auditing process. The adhesion to the program is voluntary and farmers who intend to adopt conservation management practices on their lands (with a focus on the conservation of soil and water) may participate. Eligible projects include the construction of terraces and infiltration basins, realignment of roads, restoration and protection of springs, reforestation of permanent preservation areas and legal reserves, environmental sanitation, among others.

The program is aligned with the so-called principle of provider-recipient, already widely adopted in water resources management. It provides water users who generate positive externalities with a bonus, as recognition that their practices benefit other stakeholders as well. The logic behind it is inversely analogous to the well-known 'polluter-pays' principle.

In practice, the remuneration given to farmers is always proportional to the value of the environmental services to be provided, which is estimated based on a prior inspection on the property. In order to be awarded the mark 'Water Producer', the PSA projects should follow a series of conditions and guidelines established by Brazilian National Water Agency, including the implementation of a system for monitoring results; establishment of partnerships; provision of technical assistance to the farmers involved; employment of sustainable production practices; and adoption of the watershed basin as a planning unit.

The first PSA project developed within the scope of the Water Producer Program was implemented in the city of Extrema, in Minas Gerais state. Entitled 'Conservador das Águas', it was launched nine years ago by the National Water Agency together with Sabesp (Sanitation Company of São Paulo State), IEF (State Institute of Forests), Government of Minas Gerais, SOS Mata Atlântica, Conservation International, The Nature Conservancy and Valor Natural. The project targets small rural producers from seven micro-basins and compensates them with an average remuneration of BRL 148 per hectare per year.[1] At least nine billion people living in Sao Paulo have already been direct beneficiaries of the project, particularly those that use water from the *Cantareira* Water System.

This was a pioneer experience with great success that stimulated the dissemination of the program across the country and received the '2012 Dubai International Award for Best Practices to Improve the Living Conditions' (sponsored by the UN Program for Human Settlements in partnership with the Municipality of Dubai). To date, ANA participates in 20 similar PSA projects in (...) different states (ANA, 2015).

A more recent initiative that also deserves to be highlighted was implemented by the Watershed Committee of Piracicaba-Capivari-Jundiaí. Their total area is 15,320 km^2, of which approximately 92% are located in São Paulo state (the remaining share is in Minas Gerais state). The basins cover 72 municipalities in a highly urbanized region, inhabited by more than five million people. According to the Committee's report (PCJ, 2003), water demand for the urban area is 17,300 l/s; industrial use and rural use account for 14,500 l/s and 9,100 l/s, respectively. About 31,000 l/s are required to supply half of the Metropolitan Region of São Paulo (or nine million people, approximately) and the daily pollution load is huge (i.e., 157 tDBO of domestic sewage plus 83 tDBO of industrial wastewater).

The project remunerates three practices directed to water quality and quantity: avoided deforestation, restoration of permanent protection areas (APP) and soil conservation measures for erosion prevention. The average remuneration received by

[1] One US dollar corresponds to approximately BRL 1.96 ("Brazilian Real").

each of the projects' participants is about BRL 150–200 per hectare per year, issued in biannual installments. Funds for the project realization come from payments for water use by medium and large farmers, as well as industries and companies of water supply. That includes the amount related to the PSA contracts, which are initially set up for five years and then renewed. Technical and financial support is also provided for the purchase of seeds, fencing and field operations.

Besides ANA and The Nature Conservancy, other partners (including State Secretariat for Environment, World Wildlife Foundation, Banco do Brasil Foundation, CATI and the Watersheds Committee) meet quarterly in order to deliberate not only on administrative issues but also on strategic matters. The idea is to ensure close collaboration from different stakeholders in a joint management. According to the project's report (TNC, 2010), once rural producers potentially interested in participating in the project are identified, technicians go to their properties to map the areas where reforestation and conservation measures must be undertaken. A work plan is then elaborated and submitted to the so-called Project Management Unit (formed by representative members of each partner organization) which defines remuneration values to be given. If the producer accepts the offer, a contract will be generated and signed by both parties. Target-areas mentioned in the projects are then inspected and monitored regularly with the purpose of verifying the implementation of the suggested activities, according to the proposed schedule.

In 2013, twenty seven properties benefited from it. Beneficiaries are small farmers who mostly live from agriculture (a few of them have other jobs in the city). These properties are located in two micro-basins—Moinho and Cancan—in the municipalities of Nazaré Paulista and Joanópolis.

Africa: Consortium Approach for Research and Diffusion of Vertisol Technologies in Ethiopia

Vertisols cover an area of about 12.5 million hectares in Ethiopia, out of which about eight million hectares are in the highland areas that receive sufficient rainfall. These soils have relatively good inherent fertility, but their productivity is constrained by their physical and hydrological properties, manifested by their hardness when dry and their stickiness when wet (Teklu et al., 2004). Their very slow internal drainage, with infiltration rates between 2.5 and 6.0 cm/day leads to waterlogging as result of which early planting under the traditional system is not an option. Consequently, the soils remain bare during the peak rainy season and occasional tillage operations make the soil permanently vulnerable to erosion. This does not only lead to deteriorated soil quality, but also reduces the actual growing period and hence the land and water productivity. It is believed that if half of the potential presented by these soils is achieved, it would ensure food security for the country.

Surface drainage is identified as an entry point for improving productivity of vertisols in Ethiopia. Broad Bed and Furrow (BBF) landform that is constructed by low-cost, animal-drawn implements has been identified as one of the ways to facilitate adequate drainage of these soils to enable farmers to plant early in the main rainy season. The excess rainwater can also be stored in ponds or reservoirs, and used to provide irrigation water for a second crop. BBF allows farmers to establish a first crop early in the growing season and to obtain higher and more stable yields; and the harvest of the first crop allows having a second crop using supplementary irrigation using the harvested water to facilitate germination and establishment of the crop.

At ICRISAT in India, research on vertisol technologies as a package of options started in 1974. A major lesson from this initial work was that BBF was mainly used by farmers with drainage problems in their fields (Joshi et al., 2002). Thus, subsequent research by ICRISAT on vertisols was designed to target eco-regions with the specific problem of waterlogging. Part of this was involvement in the application of watersheds concept under the ILRI-EIAR-ICRISAT-Alemaya University Joint Vertisol Project (JVP).

The Joint Vertisol Project: Evolution of Strategic Partnerships and Actor Roles

In the early 70's, priority NRM research at ICRISAT provided tools to improve surface drainage of vertisols. By the mid 70's, ICRISAT developed an animal-drawn tool carrier to form broad beds and furrows in India (El-Swaify et al., 1985). However, the implement was not readily accepted because the local zebu oxen could not produce the required draught power and the tool carrier was expensive for the subsistence farmers in India. This inapplicability to India was also experienced in Ethiopia and presented another research challenge to the team. ILCA and other national institutions collaborated in modifying BBM based on the traditional local plough (ILCA, 1990; Astatke and Kelemu, 1993).

The Joint Vertisol Project (JVP) was established with the major objective of generating economically viable technology suitable for better utilizing and conserving the production potential of the Vertisols in Ethiopia. It should be noted the BBF system was in use in Ethiopia since antiquity, but there was no tool developed for preparing the landform. Since the process of carrying out the operation by hand was tedious, development of the BBM in the late 1980s was at the core of the JVP. ILRI (formerly known as ILCA in Ethiopia) and ICRISAT were two of the five collaborating institutions for this project. A proposal was prepared by ILCA and ICRISAT in 1984 for a joint initiative with involvement of the Ethiopian government to improve Vertisol management. Figure 2.1 presents the innovation history under JVP starting in 1985 with submission of implementation documents to the Ministry of Agriculture (MoA), Institute of Agricultural Research (IAR, now EIAR), and Alemaya University of Agricultural (AUA, now Haramaya University) resulting in the official formation of JVP in 1986. The Advisory Committee (AC) and Technical Committee (TC) were constituted in March 1986. The AC considered key policy issues associated with the project and was in charge of the overall monitoring of the project. The TC comprised specialist personnel from the participating agencies and was responsible for actual project implementation. Jutzi et al. (1988) detail the functions of these two committees that ensured a high degree of collaboration between project partners. In the initial phase (1986–1989) collaboration was effected between international research centers, i.e., ILCA, ICRISAT and IBSRAM. Ethiopian national institutions participated without

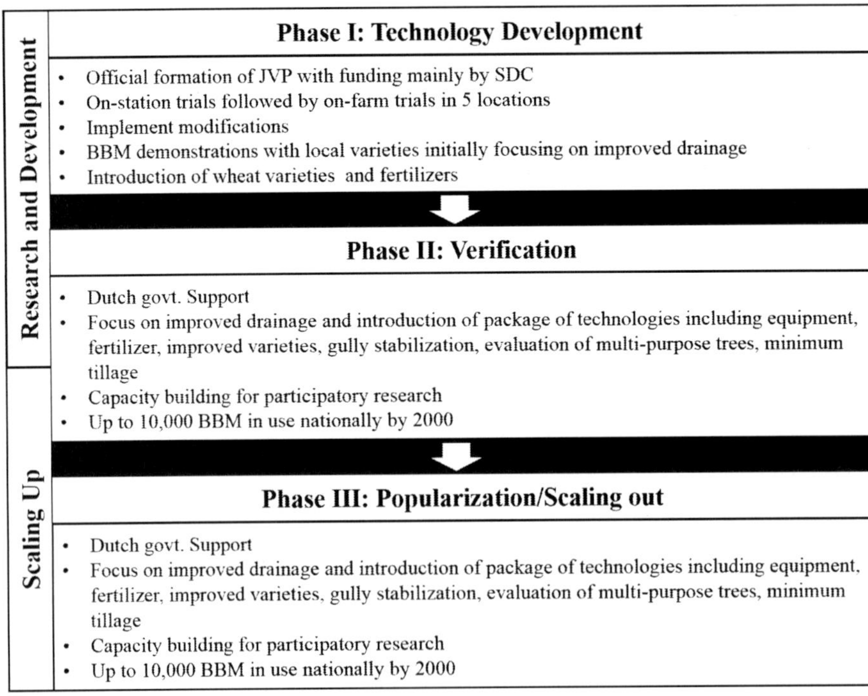

Figure 2.1. Innovation History of Vertisol Management Technologies.

direct funding at first. The five-year project (1986–1991) was mostly funded by the Swiss Development Co-operation (SDC) with additional support from Oxfam, CARITAS Switzerland and the Norwegian government.

Researchers involved in the JVP recognized the positive role farmers play as partners in the technology development process, so involved farmers in the research process used the farming systems research (FSR) approach (Amede et al., 2004). After on-station trials, the BBM package was tested by JVP on a few farms at seven Vertisol sites (Ginchi, Chefe Donsa, Degollo, Sheno, Adet, Abobo and Melka Werer) during 1986–1989. During this technology development phase, five sites were selected to verify the technical feasibility of the BBM, and to disseminate the technology to non-participating farmers and evaluate its technical performance. In 1986, the project initiated on-farm research with 56 individual farmers and three producer cooperatives in Debre Zeit, Dogollo (South Wollo) and Enewari (North Shewa), followed by Dejene in 1987 and Ginchi in 1988; and in the fifth year the number of participating farmers reached 158 along with 10 producer cooperatives. In the initial phase of the project, the problem was not identified by farmers, but rather it was a technology-driven, i.e., Vertisol drainage. Additional requirements were identified by the farmers at later stages, and the researchers responded by initiating fertilizer and varietal trials with various crops. The development of a suitable moldboard plough as a replacement for the 'Maresha' that would improve technical efficiency of handling Vertisols and enable the fulfillment of the complex socioeconomic criteria of farmers remained a priority for the project researchers. Amede et al. (2004) describes the continuous modifications and improvements on the implement taking into account farmers' feedback and suggestions.

Between 1990 and 1995, on-farm research continued at three sites (i.e., Inewari, Hidi, Ginchi) with a particular focus on the adoption behavior of farmers (Jabbar et al., 2003). In 1991, technical committee membership was limited to active participants based on the recommendations by the external review panel (comprised of AUA, IAR ILCA, ICRISAT and MoA). ICRISAT was fully engaged in 1992 and continued its active role in the project from the second phase of the JVP—commencing in 1994 with funds provided by the governments of the Netherlands and Ethiopia, and focusing the reduction of waterlogging and the introduction of a technological package (which included improved varieties and use of fertilizers).

In the first phase of Dutch government funding of JVP (from 1994 to 1995 extended up to mid-1997), the Ministry of Agriculture was not part of the consortium of institutions to implement JVP. However, its inclusion in the second phase (November 1997 to 2000) was a strategic decision and an important factor that contributed significantly in the large-scale adoption of BBM technology. This is because the Ministry of Agriculture is the key organization responsible for the extension of agricultural technologies in Ethiopia. It is at this phase of the JVP research for development continuum that the critical realization that the watersheds concept and its application was vital for lifting the farm level constraints faced by farmers in the vertisol areas of Ethiopia. Thus, the technology dissemination phase from 1993 to 2000 primarily used a watershed approach (Amede et al., 2004). The innovation at this stage was to facilitate dissemination and uptake through a coordinated and integrated watershed management approach rather than individual plots. Governance principles emerged including group collective action to manage upstream-downstream externalities as highlighted, e.g., by Kerr (2007). In this case, efforts were made to minimize conflicts among farmers on the upper and lower slopes, and to reduce land degradation caused by the water drained from the upper slopes. Technologies that accompanied the BBM included wheat varieties, fertilizers, gully stabilization, evaluation of multi-purpose trees and minimum tillage.

The watershed activities started in mid April 1998 at Chefe Donsa by identifying a watershed site adjacent to the EARO research site. The donor advocated a focus on Client-Oriented Research (COR) and formation of Farmer Research Groups (FRGs) thus driving the project to shift from the original plan to a highly participatory, feedback-oriented approach, whereby primarily priorities identified by the farmers (such as crop varieties and fertilizer) were implemented. Awareness was also created through workshops, short-courses, conferences and field days on COR approaches led by the international center for development-oriented research in agriculture (ICRA) and Royal tropical institute (KIT).

ICRISAT scientists were fully engaged in the adaptation of the watersheds concept and engineering design and layout of the BBF system for improved drainage. They also provided training on monitoring run-off and soil loss, and provided expertise in agro-climatology, soil physics, soil fertility and land and water management. ICRISAT financed and conducted training of selected technicians and seconded a scientist of JVP to work onsite on watershed management. Senior scientists also attended regular technical committee meetings. While ICRISAT innovated on the application of the watershed concept, ILCA was involved in the design of the BBM and provided training in the manufacture of animal drawn surface drainage implements to members of national partner organizations. The institute conducted research on draught animals, implements for soil, water and crop management, and water harvesting, crop-livestock interactions, crop residue utilization and on-farm technology evaluation. IAR on the other hand, was involved in nutrient management (nitrogen), cropping systems and standardization of animal drawn implement testing, while AUA was engaged in nutrient management (phosphorus), and cropping systems.

Key Drivers of Diffusion of Vertisol Technologies after JVP

In the expansion phase, since 2001, there have been demonstrations and scaling up efforts mainly by extension workers in the Ministry of Agriculture and Rural Development. Programs instituted during emergency regimes, like drought situations, advanced the involvement of agencies like Sasakawa Global 2000 and the Jimmy Carter Foundation to promote the dissemination of packages of improved technologies on-the-shelf starting early 1990s. EIAR is also involved in the actual scaling out, although its mandate is limited to pre-scaling out activities. For instance, the institute has collaborated with Sasakawa Global 2000 in promoting a package that includes small-scale irrigation using small ponds in the villages as well as treadle pumps and growing of high value crops. ILRI has also maintained interest in BBF technology and has published two separate ex-post impact assessment reports (Rutherford et al., 2001; Rutherford, 2008). Research is underway by EAIR and CIMMYT on variety selection for teff and wheat specifically for vertisols.

The enhanced financial stability of the Ethiopian government and sustained substantial support, through extension, subsidies, clustering of farmers, training and capacity building seems to be one of the key drivers of the rapid uptake of the BBMs in recent years. The Ethiopian government has been actively involved in promoting the adoption and use of the BBM technology package to reach its food production targets. Government support has been in the form of one time BBM price subsidies (farmers buy it at 50 birr[2] in Oromia region and 90 birr in the Amhara region—instead of the market price of 200 birr—from district agricultural offices), increased access to credit, and increased training. The government is also supporting youth micro-enterprises in the villages through training and provision of loans for the establishment of workshops that produce and continuously adapt the BBM. These co-operatives are creating employment for the youth but there are no proper systems to regulate the standards of the equipment manufactured. Nevertheless, besides quality and broader impacts of the intervention, the government is interested in indicators of how many youths have been employed. The equipment is then purchased by the government and distributed at subsidized rates through district agricultural research and development offices. BBMs in the Amhara region were being built by local Technical and Vocational Education and Training Centre (TVET) graduates who formed metal workshop groups using financial assistance from the government (Rutherford, 2008). In the Oromia

[2] One US dollar corresponds to approximately 16 birr.

region, the government is focusing its efforts on micro-enterprises (i.e., Urban Youth groups) to supply and distribute the BBM. Exchange visits have also been organized for farmers between the Amhara and Oromia regions including national field days on double cropping where farmers harvest the first crop grown on BBF during the main rainy season and plant the second one often on the same day to grow on residual soil moisture.

BBF technology has been taken up in new areas like the Modjo district, on the escarpment of the Rift Valley. Follow up visits of farmers groups seem to suggest that the concept of collective action could prove to be successful. For instance, some form of collective action was noted in the area where about 60 farmers have come together to multiply seed and market it. Generally, farmers met both at Ginchi and Modjo expressed satisfaction with the BBF systems by specifically mentioning that they felt the problem of water logging was solved. They reported benefits in terms of improved drainage, higher crop yields and possibility of double cropping.

One of the policies used by the government to encourage use of the BBM technology package (TP) was to set target quotas for each region, zone and Woreda with respect to the number of people receiving training (government personnel and farmers), the quantity of package inputs distributed, and the area of land drained using the BBM. Unlike the situation in 1993/94, in 2008 there was relatively little traditional non-governmental agency support for extension of the BBM apart from some localized activity of Sasakawa-Global 2000. The Rural Capacity Building Project funded by the World Bank and CIDA has also recently supported the distribution of the BBM in its operational areas. Yet in the intervening period, when there was little or no government nor NGO support for extension, there appears to have been minimal use of the BBM TP let alone 'spontaneous' adoption. This finding has serious policy implications, as it suggests that under current economic and market conditions, widespread adoption and sustained use of the BBM TP will not occur without government support. On a national scale, BBM TP adoption and impact on welfare remain relatively low, with approximately 100,000 farmers using the TP on 63,000 hectares (Rutherford, 2008). By the end of 2010, about 5,000 BBMs have been distributed in the Ginchi district alone.

Discussion and Conclusions

The two case studies show that, apart from all the complexity surrounding water, climate chance and agriculture, good ideas may yield good results even in the absence of major financial investments. Besides, agriculture's potential to enhance water quality and availability and contribute to climate change mitigation should not be overlooked.

The implementation of the Water Producer Program was the key in disseminating and consolidating PSA schemes within the context of environmental resources management in Brazil. Even though bureaucratic issues posed an obstacle to the participation of some farmers and ended up limiting the program's comprehensiveness, it served as the kick-off for other initiatives that together might reach a much larger scale. Moreover, the fact that the project's funds are ensured by law (and thus cannot be interrupted suddenly or at random) promotes the efficient and fair use of financial resources as well as the stability of the payments. Both elements are crucial for the success of any PSA scheme, since producers feel more secure being involved and investing in long-term sustainability practices throughout the duration of their contracts.

The case of vertisol management in Ethiopia is exemplary for how IARCs and their partners can draw lessons from location-specific, applied research and utilize the knowledge in other regions. The experiences under JVP and later initiatives by the government and other NGOs after the end of the project reiterates the complexity involved in research and development, especially for natural resource management technologies. Even though the target eco-region and required interventions were properly identified based on past research, several challenges were faced in implementation. All the same, a broad spectrum of stakeholders including farmers, private sector, policy makers, NGOs and researchers were involved and continuously adapted their approaches in response to challenges. Some of the major lessons learnt in the innovation process include: (1) the importance of client orientation to match interventions with actual problems of the target beneficiaries and the resources and inputs available to them; (2) the need for sustained efforts through demonstrations and training to manage perceived and actual risks; (3) the need for encouraging changes in attitudes and practices; (4) the importance of proper leadership and co-ordination to champion for change and to facilitate multi-stakeholder partnerships; and (5) the importance of synchronizing research approaches with the government development strategy and priorities. A major milestone that enhances prospects for higher adoption is the governments' prioritization of vertisol management and its institutionalization under the new Ethiopia Agricultural Transformation Plan (EATP) to tackle food security. This underlines the importance of convincing policy makers as a key element in the exit strategy to ensure sustainability.

More than intrinsically related, water, climate change and agriculture are interdependent and sustain each other. The impacts that climate change might have on the availability of water for farming activities, as well as the effects that traditional agriculture might have on climate change and water availability must be addressed with absolutely urgency.

Further research on ecosystems' resilience must be conducted in order to generate technical subsidies for the implementation of successful adaptation and mitigation policies which are truly able to reduce vulnerability and risk. However, this is a very complex task that implies decisions far beyond the technical aspects of resources scarcity and management, such as the trade-off between efficacy, efficiency and equity.

Articulation between different governance instances and stakeholders' direct involvement in decision-making processes are two conditions for ensuring success in policy design. Unless social, economic, cultural and political specificities of each region are well understood, any regulatory approach—be it based on command-and-control mechanisms and/or economic incentives—will hardly function. After all, it is important to enhance countries' capacity at the local level without neglecting the global context in which they are inserted.

References

Allan, J.A. 1998. Virtual water: a strategic resource. Global solutions to regional deficits. Groundwater 36(4): 545–546.

Amede Tilahun, Dauro Daniel, Jonfa Ejigu and Seyoum Legesse. 2004. Prospects and challenges of participatory research in natural resource management: The case of the Joint Vertisol Project (JVP). *In*: Amede, T., Assefa, H. and Stroud, A. (eds.). Participatory Research in Action: Ethiopian Experiences. AHI and EARO. Ethiopia. 126 pp.

ANA - Water National Agency http://produtordeagua.ana.gov.br//Accessed February 20th, 2015.

Assad, E. 2007. Mudanças Climáticas Globais e a Agricultura no Brasil. Revista Multiciência, ed. 8. Campinas.

Assad, E. and Pinto, H. 2008. Aquecimento Global e a Nova Geografia Agrícola no Brasil. Embrapa Agropecuária e Cepeagri/Unicamp. São Paulo.

Astatke, A. and Kelemu, F. 1993. Modifying the traditional plough maresha for better management of Vertisols. pp. 85–101. *In*: Mamo, T., Astatke, A., Srivastava, K.L. and Dibabe, Y.A. (eds.). Improved Management of Vertisols for Sustainable Crop-Livestock Production in the Ethiopian Highlands. Synthesis Report 1986–1992. Technical Committee of the Joint Vertisol Project. Addis Ababa.

Cordeiro, Luiz Adriano Maia; Assad, Eduardo Delgado; Franchini, Júlio Cezar; Sá, João Carlos de Moraes; Landers, John Nicholas; Amado, Telmo Jorge Carneiro; Rodrigues, Renato de Aragão; Roloff, Glaucio; Bley Júnior, Cícero; Almeida, Herlon Goelzer; Mozzer, Gustavo Barbosa; Balbino, Luiz Carlos; Galerani, Paulo Roberto; Evangelista, Balbino Antônio; Pellegrino, Giampaolo Q. M.; Thiago de Araújo; Amaral,Denise Deckers; Ramos, Elvison; Mello, Ivo and Ralisch, Ricardo. 2012. Plano ABC—O Aquecimento Global e a Agricultura de Baixa Emissão de Carbono. Ministério da Agricultura, Pecuária e Abastecimento—MAPA, Empresa Brasileira de Pesquisa Agropecuária—EMBRAPA, Federação Brasileira de Plantio Direto na Palha—FEBRAPDP. Brasília.

El-Swaify, S.A., Pathak, P., Rego, T.J. and Singh, S. 1985. Soil management for optimised productivity under rained conditions in the semi-arid tropics. pp. 1–64. *In*: Stewart, B.A. (ed.). Advances in Soil Science, Vol. 1. Springer-Verlag, New York.

FAO. 2003. Review of world water resources by country—Water Reports n. 23. Available at: http://www.fao.org/docrep/005/Y4473E/y4473e00.htm#Contents.

FAO, LEAD. 2006. Livestock's Long Shadow—Environmental Issues and Options. Food and Agriculture Organization of the United Nations & Livestock, Environment and Development Initiative. Rome.

FAO Aquastat. 2013. Global map of irrigation areas. Rome. Available at: http://www.fao.org/nr/water/aquastat/irrigationmap/index10.stm.

Gleick, P. (ed.). 1993. Water in Crisis: A Guide to the World's Fresh Water Resources. Oxford University Press, New York.

Gersfelt, B. 2007. Food Policy for Developing Countries: The Role of Government in the Gloval Food System—Case Study #8-4: Allocating Irrigation Water in Egypt. Edited by Cornell University, New York. 1 pp.

Hoekstra, Arjen Y. and Ashok K. Chapagain. 2008. Globalization of Water: Sharing the Planet's Freshwater Resources. ISBN: 978-1-4051-6335-4, 220 pages, December 2007, Wiley-Blackwell. Available at: http://www.waterfootprint.org/?page=files/GlobWat_page.

IFPRI. 2009. Climate change—Impact on Agriculture and Costs of Adaptation—Executive summary. International Food Policy Research Institute. Washington. vii pp.

ILCA. 1990. Annual Research Report. ILCA. International Livestock Centre for Africa. Addis Ababa.

IPCC. 2007. Fourth Assessment Report: Climate Change 2007. UNFCCC. Available at: http://www.ipcc.ch/publications_and_data/publications_and_data_reports.shtml#1.

Jabbar, M.A., Mohamed Saleem, M.A., Gebreselassie, S. and Beyene, H. 2003. Role of knowledge in the adoption of new agricultural technologies: an approach and an application. Int. J. Agricultural Resources, Governance and Ecology 2(3/4): 312–327.

Joshi, P.K., Shiyani, R.L., Bantilan, M.C.S., Pathak, P. and Nageswara Rao, G.D. 2002. Impact of vertisol technology in India. (In En. Summaries in En, Fr.) Impact Series n. 10. Patancheru 502 324, Andhra Pradesh, India: International Crops Research Institute for the Semi-Arid Tropics. 40 pp. ISBN 92-9066-453-3. Order code ISE 010.

Jutzi, S., Anderson, F.M. and Abiye Astatke. 1986. Low-cost modifications of the traditional Ethiopian tine plough for land shaping and surface drainage of heavy clay soils: Preliminary results from on-

farm verification trials. Paper presented at the Second West African Integrated Livestock Systems Networkshop held in Freetown, Sierra Leone, 19 to 25 September 1986.
Jutzi, S.C., Haque, I., McIntire, J. and Stares, J.E.S. (eds.). 1988. Management of Vertisols in sub-Saharan Africa. Proceedings of a conference held at ILCA, Addis Ababa, Ethiopia, 31 August–4 September 1987. ILCA. Addis Ababa.
Kamanda, J.O. and Bantilan, M.C.S. 2010. The Strategic Potential of Applied Research: Developing International Public Goods from Development-oriented Projects. Working Paper Series no. 26. Patancheru 502 324, Andhra Pradesh, India: International Crops Research Institute for the Semi-Arid Tropics. 36 pp.
Kerr, John. 2007. Watershed management: lessons from common property theory. Int. J. Comm. (ISO 4) Vol. 1, October 2007: 89–109.
Molden, D. (ed.). 2007. Water for Food, Water for Life—A Comprehensive Assessment of Water Management in Agriculture. International Water Management Institute. Earthscan Publisher, London.
OECD. 2013. Water Quality and Agriculture: Meeting the Policy Change—Key Messages and Executive Summary. OECD Publisher. 9 pp.
PCJ. 2003. Relatório Síntese das Bacias Hidrográficas dos Rios Piracicaba, Capivaria e Jundiaí—Situação dos Recursos Hídricos 2002/2003. Available at http://www.comitepcj.sp.gov.br/download/RS/RS-02-03_Relatorio-Sintese.pdf.
Ricardo, B. and Campalini, M. 2007. Almanaque Brasil Socioambiental 2008—Uma Nova Perspectiva para Entender a Situação do Brasil e a Nossa Contribuição para a Crise Planetária. Instituto Socioambiental—ISA. São Paulo. 317 pp.
Rutherford, A.S., Odero, A.N. and Kruska, R.L. 2001. The role of the broadbed maker plough in Ethiopian farming systems: An ex post impact assessment study. ILRI Impact Assessment Series 7. ILRI (International Livestock Research Institute), Nairobi. 156 pp.
Rutherford, A.S. 2008. Broad bed maker technology package innovations in Ethiopian farming systems: An ex post impact assessment. Research Report 20. ILRI (International Livestock Research Institute), Nairobi. 89 pp.
Teklu, E., Assefa, G. and Stahr, K. 2004. Land preparation methods efficiency on the highland Vertisols of Ethiopia. Irrigation and Drainage 53: 69–75.
TNC. 2010. Boletim do Produtor de Água no PCJ—Programa Produtor de Água, ed. 2. The Nature Conservancy Brasil.
Turral, H., Burke, J. and Faurès, J. 2011. Climate Change, Water and Food Security. FAO Water Reports n. 36. Food and Agriculture Organization of the United Nations. Rome.
UNEP. 2011. Towards a Green Economy: Pathways to Sustainable Development and Poverty Eradication. United Nations Environmental Program Report. 14 pp. Available at: www.unep.org/greeneconomy.
UNESCO. 2003. United Nations World Water Development Report—WWDR 1: Water for People, Water for Life. United Nations Educational, Scientific and Cultural Organization. Berghann Books.
WHO, UNICEF. 2008. Joint Monitoring Programme (JMP) for Water Supply and Sanitation. World Health Organization and United Nations Children's Fund. 8,13 pp.
WFN. 2013. The Water Footprint Network 2013. Available at: http://www.waterfootprint.org/?page=files/VirtualWaterFlows (last access: March 2013).
World Bank. 2007. World Development Report 2008: Agriculture for Development. The World Bank. Washington.
WWDR. 2012. Managing Water under Uncertainty and Risk. The United Nations Water Development Report WWDR4—Facts & Figures. World Water Assessment Programme. 2 pp.

3

Global Research Alliance on Agricultural Greenhouse Gases

Alan J. Franzluebbers,[1,*] *Denis Angers,*[2] *Harry Clark,*[3] *Fiona Ehrhardt,*[4] *Peter Grace,*[5] *Ladislau Martin-Neto,*[6] *Brian McConkey,*[7] *Leann Palmer,*[8] *Sylvie Recous,*[9] *Renato de Aragão Ribeiro Rodrigues,*[10] *Alvaro Roel,*[11] *Martin Scholten,*[12] *Steven Shafer,*[13] *Bill Slattery,*[14] *Jean-Francois Soussana,*[15] *Jan Verhagen,*[16] *Kazuyuki Yagi*[17] and *Gonzalo Zorrilla*[18]

Background

The Global Research Alliance (GRA) on Agricultural Greenhouse Gases was established in the margins of the Conference of Parties (COP15) in Copenhagen, Denmark on 16 December 2009. However, the idea was borne earlier in the year from discussions originating from New Zealand scientists and negotiators at other climate change conferences. With imperiled climate change negotiations of the United Nations Framework on Climate Change Convention (UNFCCC) occurring earlier in the year and culminating in Copenhagen, an agreement was signed by government ministers of 21 countries to form the GRA and create a positive step forward for the agricultural sector, which previously was left without much voice in the UNFCCC negotiations. The 21 countries agreeing to participate in the GRA were Australia, Canada, Columbia, Chile, Denmark, France, Germany, Ghana, India, Ireland, Japan, Malaysia, Netherlands, New Zealand, Spain, Sweden, Switzerland, United Kingdom, United States, Uruguay, and Vietnam. The logo of the GRA is shown in Fig. 3.1.

Authors' affiliations given at the end of the chapter.

GLOBAL
RESEARCH
ALLIANCE
ON AGRICULTURAL GREENHOUSE GASES

Figure 3.1. Logo of the GRA developed following its inception in December 2009.

The GRA was formed to build upon the positive scientific understanding of greenhouse gas (GHG) emissions research within the agricultural community of many countries, as well as to bring greater visibility to the larger role that agriculture could undertake in mitigating global GHG emissions. An enhanced research pace and greater breadth of relevance were expected to be accomplished with international cooperation and investment among countries, i.e., more could be achieved by working together than working separately.

The original and continued goal of the GRA was to reduce GHG emissions intensity of agricultural production systems and increase their potential for soil C sequestration. From the Joint Ministerial Statement produced at the initiation of the GRA in December 2009, the following aims were declared:

- Agriculture, including livestock, cropping, and rice production, plays a vital role in food security, poverty reduction, and sustainable development;
- The agricultural sector is particularly vulnerable to impacts of climate change and faces significant challenges in meeting a dramatic increase in global food demand, while reducing its contribution to global GHG emissions;
- Opportunities exist to reduce agricultural GHG emissions and increase C sequestration by improving efficiency and productivity of agricultural systems through improved management practices and technologies, which can help build resilience and adaptive capacity of agricultural systems to meet the increasing demand for food in a sustainable manner;
- Underlining the need for food security, we decide to establish a GRA on agricultural GHGs to help reduce emissions intensity of agricultural production and increase it potential for soil C sequestration, thereby contributing to overall mitigation efforts;
- The GRA will seek to increase international cooperation, collaboration, and investment in both public and private research activities to:
    - Improve knowledge sharing, access to, and application by farmers of mitigation and C sequestration practices and technologies, which can also enhance productivity and resilience;
    - Promote synergies between adaptation and mitigation efforts;

- Develop science and technology needed to improve measurement and estimation of GHG emissions and C sequestration in different agricultural systems;
- Develop consistent methodological approaches for measurement and estimation of GHG emissions and C sequestration to improve research coherence and monitoring of mitigation efforts;
- Facilitate exchange of information between scientists around the world;
- Help scientists gain expertise in mitigation knowledge and technologies through new partnerships and exchange opportunities;
- Develop partnerships with farmers, farmer organizations, private sector, international and regional research institutions, foundations, and other relevant non-governmental organizations to facilitate and enhance coordination of research activities and dissemination of best practices and technologies.

These clear statements were intended to refocus stalled COP negotiations to bring countries, organizations, and individuals together in a bottom-up, voluntary network to increase international cooperation, collaboration, and investment in agricultural GHG research.

The First Senior Officials Meeting took place a few months later in Wellington, New Zealand from 7–9 April 2010. At this meeting, government officials of countries endorsing the Copenhagen Ministerial Statement gathered to create a roadmap to achieve goals for the first year. These goals included (1) development of a charter to guide the work and operations, (2) planning and undertaking of a first meeting of each of the research groups to identify research needs and priorities and develop a work plan, and (3) establishing a partnership network of non-government organization for collaboration. Work of the GRA began in earnest with various meetings of different groups throughout 2010. It may be useful to view the structure of the GRA focused on the work of three research groups and two cross-cutting groups governed by an Alliance Council and supported by and Alliance Secretariat (Fig. 3.2).

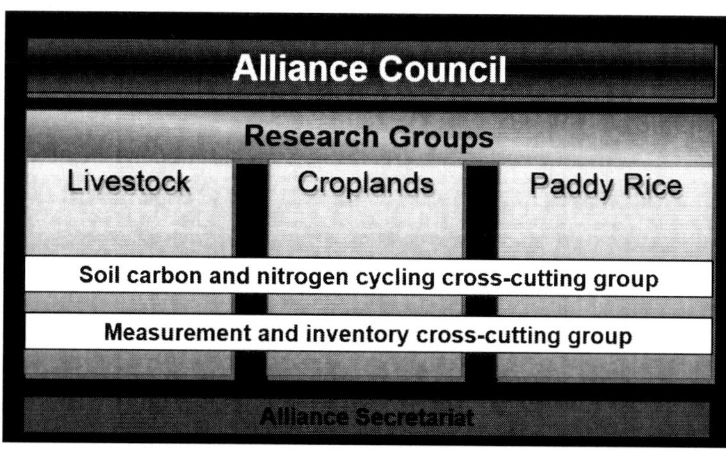

Figure 3.2. Structure of the GRA as conceived at the inaugural Ministerial Summit in Rome, Italy on 24 June 2011.

The Second Senior Officials Meeting was held in Versailles, France from 28 February to 4 March 2011. This is where each of the three research groups met for a second time and the Alliance Council hammered out issues for the development of the Alliance Charter. The Alliance Charter was signed by ministerial representatives of 31 countries at the inaugural Ministerial Summit held in Rome, Italy on 24 June 2011.

As of April 2014, a total of 40 countries belong to the GRA: Argentina, Australia, Belgium, Bolivia, Brazil, Canada, Chile, China, Colombia, Costa Rica, Denmark, Ecuador, Finland, France, Germany, Ghana, Honduras, Indonesia, Ireland, Italy, Japan, Malaysia, Mexico, Nicaragua, Netherlands, New Zealand, Norway, Panama, Peru, Philippines, Republic of Korea, Spain, Sri Lanka, Sweden, Switzerland, Thailand, United Kingdom, United States, Uruguay, and Vietnam.

Structure of the GRA

Currently, there are five groups organized around themes of croplands, livestock, paddy rice, soil carbon and nitrogen cycling, and inventories and measurement. Each group operates somewhat independently with member country representation in any or all groups. Typically, each group meets annually, although there is significant sharing of information among group leaders at quarterly teleconferences and annual meetings associated with the Alliance Council gathering. To increase collaboration and enhance research synergies, all research groups have agreed to work under a common thematic structure depicted in Fig. 3.3. This structure has two lines of activities, one based on common understanding for a member country, partner, or individual scientist and the other based on assembling the shared vision into concerted actions representing the GRA. Stocktake and inventories are used to assess the capabilities and resources available from various members and partners involved in the GRA. Putting individual research contributions together yields a more rigorous output through enhanced networks and databases. Capacity development helps under-resourced members make more significant contributions, either through education of young scientists or resourcing with shared equipment and data collection systems. By building capacity and working together, new and effective collaborative research partnerships can be achieved. Information and technology transfer are goals for bringing greater value of research endeavors to private entities and public institutions. Assembling a broad

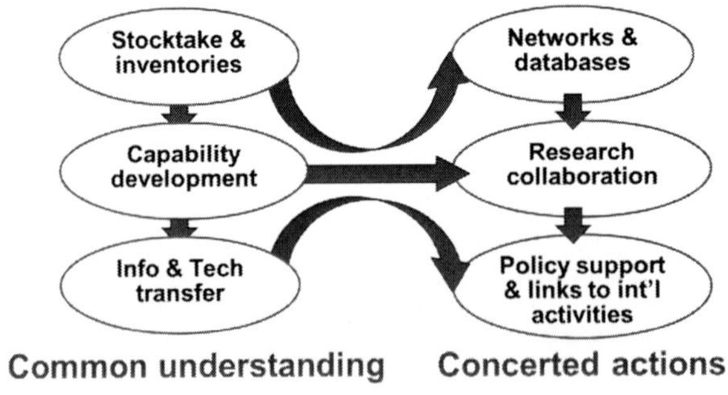

Figure 3.3. Approaches undertaken by each of the research and cross-cutting groups in the GRA.

body of research data into a working package will allow our research community to provide essential policy support with potentially limited contributions from any one country, but through this alliance can provide robust connections to research activities in other countries.

Groups have contributed to the overall goals of the GRA by forming collaborative research projects, developing research networks, creating joint funding initiatives, sharing data, transferring information and technology, networking with other organizations to leverage resources and create synergies, and developing capacity among members.

Collaborative research projects accelerate knowledge and technology development that would not have occurred without the GRA. By defining a problem collectively, regional and global solutions can be achieved rather than simply at a local level. This process avoids duplication of resources otherwise necessary if operating within individual countries. By pooling limited resources and using complementary skills, more wide-spread and robust results can be achieved. This process works if members understand the need to collaborate rather than compete.

Research networks accelerate research and science coordination in targeted areas. By creating dedicated expert communities at an international level, knowledge and technologies can be shared through rapid means of communication. Sharing information in the network leads to robust methods that are trustworthy, creating more effective mitigation assessments and development of a stronger suite of good agricultural practices. In the process, greater capability among member-country scientists is developed.

Joint funding initiatives advance opportunities for efficient and effective use of limited research funding. By fostering science excellence across national boundaries, member-country resources can be leveraged and a critical mass of scientific expertise can be assembled. Better linkages are expected among scientists and institutions to be able to solve problems quicker and with limited resources.

Sharing data facilitates a culture of information exchange. By contributing to shared datasets, critical knowledge gaps can be readily identified. Filling these data gaps can be jointly achieved within an effective network. Sharing data also encourages development of common measurement protocols and good research practices.

Transferring information and technology promotes benefits of and improved access to a greater variety of knowledge and technologies in different cultures and regions of the world. Through a shared process of harmonizing methods and approaches, producing literature reviews, assembling technical manuals, and developing good practice guides, information and technology can be transferred among scientists and to industry representatives and policy developers.

Developing capacity among members will more widely reduce GHG intensity consistent with regional economies and development goals. With facilitated training and educational opportunities, young scientists from developing countries can be engaged early in global exchanges to build careers improving emission estimates for local and regional agricultural systems of greatest interest for their respective country.

Research Groups

Croplands research group

The Croplands Research Group (CRG) is currently led by the USA and Brazil. The vision of the group is to (1) develop and support widely available decision-support tools for reducing GHG emissions intensity from croplands, thereby producing sustained or increased yields for a climate-resilient agriculture and (2) create croplands that have greater soil C sequestration.

Research emphases are on (1) GHG emissions/pathways for the three primary agricultural gases of concern, i.e., nitrous oxide (N_2O), carbon dioxide (CO_2), and methane (CH_4), (2) evaluations to understand soil condition, type and frequency of tillage, cultural inputs, inorganic and organic amendments, crop varieties, crop rotations, weather conditions, etc., and (3) technologies and management practices relevant to farmers and specific field applications.

The group is organized into three primary components: (1) quantifying net GHG emissions in cropland management systems, (2) assessing GHG emissions in agricultural peatlands and wetlands, and (3) modeling C and N emissions. The group plans to promote synergies between mitigation and adaptation throughout each of these components.

Component 1: Croplands are sensitive to climate change and are a net contributor to GHG emissions. Croplands are also diverse in time, space, and variety of crops grown around the world. There is a need to quantify the role of the large diversity of cropland management systems to reduce GHG emissions. Some of the primary management techniques that have worldwide relevance for soil C sequestration and GHG emissions are primary crop, crop rotation, cover cropping, fertilization, tillage, and residue removal. Results have been variable, but without a meta-analysis of the data, it is difficult to interpret the cause of variability (e.g., function of soil, climate, and agricultural conditions). There is a need to synthesize this information to propose a set of good agricultural practices for particular soil types, ecoregions, etc.

Research is conducted to identify the components of a cropping system that have an impact on soil C sequestration, GHG emissions, and other environmental responses, both positive and negative. Under the diversity of private, university, and federal agricultural research networks around the world, there was a recognized need to establish a searchable literature database, as well as to conduct a thorough review of the literature so that we might know better the gaps and the most fruitful strategies forward. Networking via the GRA allows the group a unique opportunity to develop a global network of experimental sites and research expertise. Data collected from the group is being shared with the C and N Cross-Cutting Group to facilitate model development and verification. There is also an urgent need to quantify indirect GHG emissions derived from agricultural activities.

Anticipated products from this component of the group are:

- Standardized/acceptable protocols and improved methods for determining soil C sequestration and GHG emissions;
- International database of existing and new research on GHG fluxes and soil C sequestration rates as affected by particular agricultural management systems;
- Synthesis of currently available experimental results around the world;
- Guidelines/best practices for minimizing GHG emissions and maximizing soil C sequestration under various climatic conditions, ecoregional delineations, and/or soil types; and
- Summary documents for use by international negotiating bodies concerned with GHG emissions, soil stewardship, and natural resource management.

Potential benefits from these shared activities will be:

- Standardized datasets and data management protocols to enhance research opportunities by various nationally led research organizations;
- Greater international cooperation so that multi-national research efforts can be expanded with funding supported for specific research goals in the GRA;
- Enhanced ecosystem services that will benefit society in general, such as climate regulation, C and N cycling, water infiltration and cycling, biodiversity enhancement, and scenic landscapes.

Component 2: Agricultural production systems are cultivated on mineral soils, peatlands, and wetlands. In some countries and regions, peatlands and wetlands constitute an important part of the available land resource. Globally, agriculture is the most common use for peatlands (Oleszuk et al., 2008). Previous research has focused on how to utilize peatlands for different purposes, including cultivation for agriculture, mainly because of their high fertility and inherent value for cultivation. However, GHG emissions from cropping systems cultivated on peatlands and wetlands can be several-fold greater than from cropping systems on mineral soils. This is sometimes due to the type of cropping and management systems, but also due to cultivation itself, with the inherent necessity for drainage of peatlands to be productive for agricultural crops. Drainage enhances decomposition of peat and leads to enhanced emission of CO_2 and N_2O, but reduced emission of CH_4 (Oleszuk et al., 2008). Current research is focused on mitigating GHG emissions from peatlands and wetlands.

Peatlands that have been cultivated for many years will subside, oftentimes leading to questions by society about their suitability for cropping and the need for restoration to a more natural state to reduce further GHG emissions. Further, climate change in boreal regions is expected to lead to thawing of permafrost, which would further increase GHG emissions from peatlands. Peatlands are also used for other purposes than cropping, suggesting that research is needed in a broader societal context.

Research is needed for a global overview/synthesis of peatland resources and GHG emissions related to agricultural activities. Although peatlands predominate in boreal regions of the world, other ecoregions have significant peatlands and wetlands as well. Therefore, soil C sequestration and GHG emission research is of interest to the global research community, especially since previous research has documented the importance of these unique soils for significant GHG emissions (Oleszuk et al., 2008; Maljanen et al., 2010). Peatland soils are sensitive to climate change, and at the same time, are important sources of net GHG emissions. Country-specific research on peatlands could, therefore, be complemented with research from a global network to broaden the research experience and develop synergies among research scientists at different institutions and working under different environmental conditions. Results from such research will be a valuable input to Components 1 and 3 of this group.

Research will focus on:

- Quantification of GHG emissions from peatlands and wetlands in different geographical regions, climate conditions, cropping systems, and management practices;
- Implementation of improved management strategies to reduce GHG emissions from cropping on peatlands, e.g., cover cropping, residue management, fertilization, crop rotation, tillage system, and drainage regime;
- Restoration methods to preserve peatlands and reduce GHG emissions;
- Recommendations for improved cultivation methods and crop management strategies on newly cultivated peatland soil brought into production (e.g., as a result of thawing of permafrost or expanding cropland for food production); and
- Development of a global network of experimental sites to conduct synergistic research on soil C sequestration and GHG emissions.

Anticipated products will be:

- Overview reports of ongoing research/status of peatlands related to GHG emissions (e.g., reports posted on GRA website will give information about regional activities, research group members, and research summaries of existing and recommended practices in member countries);
- Publications/reports on recommended practices and their impacts on reducing GHG emissions, as a result of this new joint research cooperation;
- Compilation of GHG emission datasets that will contribute to the efforts of Component 1 about GHG emissions from cropland management systems, as well as be available for use in proposed C and N modeling research in Component 3;
- Recommendations for improved technologies/good agricultural practices to restore peatlands to a state of more naturally occurring ecosystem functioning; and
- Data available for the GRA cross-cutting research team on inventories and measurement methods, through knowledge transfer, datasets, discussion notes for methods, overview of existing methods, and contributions at seminars or other discussions forums.

Potential benefits from research conducted under this component are expected to include overview and synthesis products for educating the science community, and society in general. Society can find a contact point and webpage giving regional information about status, ongoing research, and recommended practices. This can give background for policymakers seeking information about possible practices, recommendations, and actions from different regional areas. It also gives possibilities to establish contacts at different levels and to different groups in the GRA. These activities can be part of the knowledge base when both national and international strategies and regulations are developed. For science, additional benefits might include:

- Possibilities to contact other research groups for joint research;
- Harmonizing methods and guidelines for collaboration/utilization of experimental sites;
- Joint modeling investigations;
- Increased possibility to spread research information/results to a global scientific community;
- Through a link to the cross-cutting issues team on inventories and measurement methods, the scientific community will further benefit from the joint work; and
- With a strong international network bringing active research groups together, greater possibilities for securing multi-national research funding may occur.

Component 3: The focus of this component is on simulation modeling of N_2O emission and soil organic C stocks and changes, because of the importance of these two processes on the global GHG budget of cropland soils. Nitrous oxide emission is a result of two soil microbiological processes: nitrification and denitrification. These processes depend on many variables (N and organic C substrate availability, temperature, soil water content), which are controlled by climatic and soil factors, as well as agricultural management practices. Nitrous oxide emission is characterized by a high variability in space and time. Therefore, it can be difficult to measure without enormous investments in time, equipment, and labor resources. Difficulties are also present in interpreting and comparing data from different experiments. Existing models range from complex process-based models simulating the dynamic of water, solutes and microbial processes on a fine-scale (DNDC, DAYCENT, Ecosys, etc.) to simple, empirical tools based on statistical inference.

Soils are an important stock of C and changes in the soil C stock affect the net emission of CO_2 to the atmosphere. Long-term change in soil C stocks depends on the amount and nature of organic matter input and on numerous soil and climatic factors and agricultural management practices, which affect organic matter decomposition. Current understanding of how multiple interacting factors influence soil C dynamic has been embodied in a wide range of soil organic matter models (e.g., Century, RothC, AMG, etc.), which have a common theoretical basis, but which also differ according to numerous aspects (e.g., the number of organic component pools, time steps, etc.).

Since a major objective of the GRA is to identify and develop agricultural practices that improve the global GHG budget of cropping systems, this research component will help address this assessment across regions not directly sampled, but rather simulated. Both N_2O emissions and soil C dynamics are strongly influenced by agricultural management practices, although at different time scales. Responses of C and N dynamics depend on many circumstances. Simulation models are needed to decipher the relative effects of soil properties, climate, and agricultural management practices for a wide range of circumstances. Their use is however hampered by the lack of synthesis documents giving an overview of existing models and of their possible use.

Few review articles have been published about N_2O emission modeling (Chen et al., 2008). This is less the case for soil C models, probably because research in this area has been active for a longer time. It remains difficult, however, to easily select the model that best suits a specific objective in a specific set of conditions or circumstances. This is especially true for non-modeling groups. One of the key considerations in selecting an ideally suited model is the breadth and intensity of input data needed to obtain a useful output from the model. Secondly, availability of data at the national level will also be needed for national inventory purposes. Therefore, the domain of the model (i.e., field or farm level or national inventory level) needs to be clearly selected and input data collected accordingly.

A major objective of this component will be to provide synthesis products that will be helpful for scientists (modelers and experimentalists) and agricultural managers and policy makers to get an overview of existing models, their input requirements, their potential outputs, and the purposes for which they are relevant. Inter-comparison of models and benchmarking are not in the scope of the group. This will be done by the C and N cycling cross-cutting team.

Anticipated products from this component are expected as:

- List of publications using (a) N_2O emission models and (b) soil organic C models;
- Review articles describing (a) N_2O emission models and (b) soil organic C models;
- Bibliometric analysis of the worldwide scientific literature on both topics and a map of the main research groups that are active on these topics;
- Evaluation of models of (a) direct N_2O emissions and (b) soil C dynamics with information on spatial scale (e.g., laboratory, field, landscape, regional, etc.), time scale, input data requirements, main simulated processes, context and range of situations tested, purposes for which they are suitable (e.g., test of researcher hypothesis, decision support system for mitigation, regional inventories, etc.), and main related publications;
- Short list of recommended models that have been widely used and tested in a wide range of situations for a particular set of conditions and purposes; and
- List of models that use a mass balance approach in considering the cycling of both C and N within the same model framework.

Potential benefits from this research component can be expected:

- For model users, an easier and better selection of models to suit specific goals;
- Facilitated access to models for a range of purposes;
- For the scientific community, experimental data interpretation;

- Soil C change and N_2O emission prediction and inventory; and
- In conditions or regions without extensive field experimentation, widespread identification of agricultural management practices that minimize N_2O emission and sequester C in soils.

Livestock research group

The Livestock Research Group (LRG) is focused on promoting ways to reduce GHG intensity of livestock production systems and increase soil carbon storage in lands supporting animal production. The LRG has identified the following vision for its livestock-related research activities:

1. Increase agricultural production with lower emissions—"Feeding the world within the carrying capacity of earth"
2. Improve global cooperation in research and technology—"Accelerate/strengthen knowledge and technology development that would not happen without the GRA"
3. Work with farmers and partners to provide greater knowledge of agricultural production systems—"Develop relevant mitigation options and strengthen productivity and resilience of food systems".

The LRG is co-chaired by New Zealand and the Netherlands. The LRG has an ambitious work plan that is regularly updated as new activities are identified and existing activities progress or are completed.

The LRG has two working groups: (1) ruminant working group and (2) non-ruminant working group. Both working groups have the following common tasks:

- Share and exchange information and methodologies, including through the development of best practice guides and technical manuals;
- Build research capacity and identify possibilities for joint research/concerted action to measure, monitor and mitigate GHG emissions and carbon sequestration, including through the formation of dedicated research networks;
- Connect and develop synergies with other institutions or individual scientists that are, or could be, relevant to the work of the LRG, including the dissemination of solutions to farmers.

Stock-take of research activities: When the GRA was first established, it was agreed that all participating countries should complete a stock-take of relevant domestic research projects and programs. The stock-take of research activities is a snapshot in time of the current investment in agricultural GHG emissions research in GRA-member countries. This was intended to enable the development of a collective understanding of individual countries' current research activities and priorities for agricultural GHG emissions research. Analysis of countries' data for the stock-take helped the LRG shape its work program. A summary of this analysis was presented to the Non-CO_2 Greenhouse Gas Symposium (NCGG-6) in November 2011 in Amsterdam, the Netherlands. The stock-take will be updated every 3–5 years. However, new-member countries will be asked to complete the stock-take. A less formal, open network of researchers, policy makers, industry representatives and students interested in livestock emissions research is provided through the Livestock Emissions and Abatement Research Network (LEARN). The LEARN website aims to facilitate and promote interaction and dialogue between individuals doing work in this space—for more information, please see http://www.livestockemissions.net/.

Research networks and databases: Dedicated research networks and databases have been identified by the LRG as useful tools to accelerate research efforts and achieve outcomes that no country could achieve on its own.

The *Rumen Microbial Genomics Network* was initiated in New Zealand in 2011 through sponsorship from the New Zealand government. This is a virtual global network of researchers who collaborate to speed up the development of rumen microbial genomics approaches for improved animal production and reducing methane emissions from ruminant livestock. The network has developed a framework for the international coordination of rumen microbial genomics projects, allowing information to easily be accessed by researchers in the field via the website and newsletters. The website is well-linked to enable easy access to methods, genome sequence and metagenome data relevant to the rumen microbial community.

The *Animal Selection, Genetics and Genomics Network* is being led by New Zealand and Australia. The network is focused on bringing scientists together who are working in the area of reducing GHG emissions from ruminant livestock through animal selection, genetics and genomics techniques. Through this increased sharing of experiences and data, and identification of new collaborations and connections, we can increase progress towards much-needed solutions.

The *Feed and Nutrition Network and Database* were established in 2012. Feed and nutrition are some of the most intensively researched areas in relation to agricultural GHG mitigation and this research will benefit greatly from increased international collaboration, sharing of information, and coordination of further research.

The *Manure Management Network* focuses on emissions of methane and nitrous oxide from manure, as it contributes to GHG emissions from livestock worldwide. Research on mitigation measures for these emissions is urgent and would benefit greatly from increased international collaboration. This collaboration will be facilitated by the establishment of an international network on manure management relating to GHG emissions from livestock, which is being led by the Netherlands and Vietnam. The network will help increase information-sharing and coordination of international research. The network is focused on bringing scientists together who are working in the area of reducing GHG emissions from livestock through improvement of manure management. Some initial areas of focus for the network are:

- Develop a best practice guide to measure emissions from manure in all stages of the manure chain;
- Make a position paper and leaflet to be used for external communication dealing with goals, role, position/boundaries, etc.;
- Make a shopping list on practical mitigation options for farmers and policy.

One of the themes in the Global Agenda of Action of the Livestock Dialogue of FAO is reduced discharge of animal manure. The goal of this theme is reducing nutrient overload and GHG emissions through cost effective recycling and recovery of nutrients and energy contained in animal manure. The network agreed to cooperate with the Livestock Dialogue on this theme and to develop and execute a new joint working program, i.e., the Manure Management Improvement Program.

The *Animal Health and Greenhouse Gas Emissions Intensity Network* recognizes there is broad consensus amongst experts and stakeholders that GHG emissions intensity from livestock farming could be reduced through efficiency and production gains resulting from improved livestock health. In light of this, the UK is leading the establishment of a network aiming to identify and explore synergies between animal health and GHG emissions intensity and to facilitate the interaction of researchers from relevant research communities. The objectives of the network are to:

1. Share information on current and planned funding activities in the field of animal health and GHG emissions intensity, so as to avoid duplication of effort, identify gaps and help focus research efforts;
2. Maintain and enhance capacity in this field of research, including the ability of practitioners from GHG emissions intensity, animal health and other relevant fields to interact;
3. Encourage and facilitate a joined-up approach from fundamental science to strategic and applied research and research-into-use while avoiding overlaps and identifying gaps and opportunities for collaboration;
4. Establish common agreement on priority issues and explore funding opportunities to address them, including links with more traditional animal health and agricultural and rural development programs;
5. Pursue synergies with stakeholders, including STAR-IDAZ, OIE, FAO, World Bank, and rural development organizations to further strengthen global cooperation and networks.

The *Livestock Emissions Abatement Research Network* (LEARN) is an international network to facilitate researcher-to-researcher connections that will ultimately lead to the development of practical and cost effective agricultural GHG mitigation solutions. LEARN links scientists, students, policy makers and industry people together who are interested in finding agricultural GHG emissions mitigation solutions. Joining LEARN involves uploading your details and information about your research, your current research projects and future research interests. A search engine will facilitate your connections with other interested people across the world. LEARN also provides a scholarship and training fund, sponsored by the New Zealand Government, to build capability among technicians, emerging researchers and senior scientists. The LEARN fund is designed to enable researchers to acquire skills in new/unfamiliar techniques and methods in agricultural greenhouse gas emissions mitigation.

The *Grassland Research Network* focuses on land-based animal production. Rangeland covers half of the land worldwide, largely used for livestock grazing because of its vegetation cover and soil characteristics. In addition, grasslands represent more than one-fourth of the world's terrestrial surface, making the livestock sector the largest user of agricultural land. Therefore, implementing better management strategies to improve production efficiency of grazing livestock could have an important impact on global GHG emissions of agriculture. The network has the following objectives:

- Promoting best management practices related to carbon sequestration (GHG removals) at the farm level and potential of carbon sink in different farming systems;
- Improving understanding of the implications of soil carbon losses and degradation (synergies adaptation and mitigation);
- Identifying knowledge gaps and opportunities for research collaboration and dissemination between grasslands systems and rangelands systems.

The network aims to build capacities on a topic of significant interest to developing country regions, playing an important role in the range of activities pursued by the LRG. Collaboration with other research groups and partner initiatives are being explored, given the significant synergies between the goals of the network with organizations such as the International Livestock Research Institute, the Global Agenda of Action established by the FAO, and the modelling activities developed by the Soil C/N Cycling Crosscutting Group of the GRA.

Capability development and fellowships: Members of the LRG actively promote and encourage capability development through research training and development and targeted research opportunities. Two strategies have been agreed to by the Group to date:

- Regional technical capability building workshops—focused around the issues of measuring and mitigating GHG emissions from livestock systems;
- Fellowship opportunities—several countries have made fellowship opportunities available to the GRA to help build capability and capacity.

Collaborative research projects: Periodically, the LRG identifies priority areas for collaborative research projects. Countries that are interested in participating in the projects then collaborate to source funding to initiate these activities. Some projects underway include:

- Improved inventory and mitigation of GHGs in livestock production in South East Asia
- Deep sequencing the rumen microbiome
- Accelerated discovery of methanogen-specific inhibitors
- Vaccine to reduce methane emissions in ruminants
- Animal delivery of DCD in urine by provision in feeds
- Low emitting animals
- Characterizing rumen microbial diversity
- Climate change and livestock: Quantification and mitigation of methane and nitrous oxide emissions from grazing livestock systems
- Automated methane measurement: Evaluating the C-Lock system
- Technical manual on respiration chamber design

Technical information and knowledge hub: Outputs or reports produced by collaborative research projects are available at: http://www.globalresearchalliance.org/research/livestock/activities/knowledge/. A priority area of work for the LRG is the development of internationally co-authored best practice guidelines and technical manuals. These manuals will greatly assist the international research effort into livestock GHG mitigation by helping improve consistency of methodologies, techniques and practices used, and therefore the comparability of results. Updates and science reports of collaborative research projects supported by the LRG are also available on this website.

Policy support and linkages to international initiatives: Developing relationships with other institutions, organizations, and individuals working in similar areas is a critical focus for the LRG so that shared goals and areas of common interest can be identified and relevant activities worked on together. In this way, we can avoid duplication of effort and can leverage strengths and unique areas of focus. Such institutions could include regional or global scientific and capability building organizations, regional and multi-lateral funding agencies, regional and significant national research institutes, and outreach and extension agencies. The LRG is currently engaging with the following organizations:

- CGIAR Climate Change, Agriculture and Food Security Program
- European Commission Directorate-General of Research and Innovation
- European Joint Programming Initiative on Agriculture, Food Security and Climate Change
- Food and Agriculture Organization of the United Nations
- Global Agenda of Action in Support of Sustainable Livestock Sector Development
- International Livestock Research Institute
- World Bank
- Animal Task Force
- International Dairy Federation
- International Meat Secretariat

The Livestock Research Group also has links to other initiatives, often seeking to hold its meetings in conjunction with relevant conferences and events, including:

- Global Conference on Agricultural Research for Development, November 2012
- Non-CO_2 Greenhouse Gas Symposium, November 2011
- Greenhouse Gas & Animal Agriculture Conference, October 2010

Paddy rice research group

Paddy rice production causes significant CH_4 emissions in comparison to other cropping systems and, thus, GRA members envisioned a need to establish a separate research group on this globally-important crop. The Paddy Rice Research Group (PRRG) is working together to find ways to reduce the emissions intensity of paddy rice cultivation systems, while improving its overall production efficiency. Trade-offs with emissions of N_2O and changes of the quantity of C stored in paddy soils are also being considered.

The PRRG is currently co-chaired by Japan and Uruguay. In 2013, 22 countries participated in the group as members. Some non-member countries that have significant rice production are joining activities of the group as observers. The PRRG has organized five annual meetings, in Japan (2 times), France, Philippines, and Indonesia.

The work of the PRRG is focused on helping provide knowledge of source/sink processes and mitigation options to paddy rice farmers, land managers, and policy makers by looking at the impacts of water management, organic matter and fertilisers, and cultivar selection. It will also help to improve national inventories of GHG emissions from paddy rice cultivation systems.

The PRRG is conducting activities on the following topic areas:

- Standardising measurement techniques;
- Developing databases for relevant experiments and experts;
- Increasing participation of countries and partners; and
- Conducting multi-country field experiments for testing mitigation options.

Measurement techniques: Following on from the PRRG survey of measurement methodologies, the PRRG recognised the need to review and standardise measurement techniques of member countries so that experimental results may be shared and compared within the PRRG. As the first step, the PRRG collected measurement protocols and guidelines used in the countries/institutes of members to form an understanding of common protocols and reasons for any differences. Comparison of the methods identified items to be included in a standardised protocol, such as field design, chamber, gas collection, timing of gas sampling, and analysis. From these exercises, the PRRG plans to publish a manual of standardized measurement techniques with the identification of "good practice" options for each region and climate.

In addition to comparing measurement protocols, the PRRG exchanged automated measurement data for identifying the best time of the day and the frequency of measurements required to obtain representative values of CH_4 emissions by the closed chamber technique. As a result, an original scientific paper from analysis of Japanese data was published (Minamikawa et al., 2012). Additional analyses of data in tropical conditions will be conducted.

Databases: The PRRG agreed to develop databases for literature and experts that will enable members to search publication references on GHG and paddy rice research according to management system or region. Further discussion led to a decision to contribute to the database that compiles metadata from experimental sites throughout the world where GHG fluxes are monitored, based on the CRG Managing Agricultural Greenhouse Gases Network (MAGGnet) activity. This database would combine and replace the original Alliance stocktake, which the PRRG had planned to complete for rice research and the literature/expert database that was being developed. The spreadsheet for each experimental site would capture information about the experiment aims and methods, researchers and organisations involved and include specific site information (e.g., soil type, rainfall, rice cultivar).

Participation: In order to encourage the participation of scientists in other countries, as well as extending the activities and outputs of the PRRG, the PRRG co-organizes international symposia on mitigating GHG emissions from paddy fields. The first capability building workshop to focus on this outreach was for Latin-American countries, and was held in June 2013 in Montevideo, Uruguay. A symposium entitled "Mitigating Greenhouse Gas Emissions from Rice Paddy Soils" is also planned at the 20th World Congress of Soil Science to be held in Jeju, Korea from 8–13 June 2014. The PRRG continues to collaborate with research centers and programs of CGIAR (IRRI, CIAT, AfricaRice, and CCAFS). These groups directly collaborate in research projects and participate in group meetings.

Field experiments: A new research project, MIRSA (Greenhouse Gas Mitigation in Irrigated Rice Paddies in Southeast Asia), is being implemented to assess the feasibility of GHG mitigation through water saving techniques in irrigated rice. This project is being funded by the Ministry of Agriculture, Forestry and Fisheries of Japan from 2013 to 2018. Participants in the project are from Japan, Indonesia, Philippines (together with IRRI), and Vietnam. This project will assess water management options. Similar Project for North-South America are being discussed for implementation.

Soil carbon and nitrogen cross-cutting group

Scope and workplan: The soil carbon and nitrogen cycling cross-cutting group (SCNG) is currently led by France and Australia and has 20 participating countries. In close collaboration with the three research groups (Croplands, Livestock and Rice), the SCNG aims at: (i) developing improved methodologies and models for mitigation, (ii) building collective expertise on applicability of models, uncertainty and range of mitigation options and (iii) constructing a common modelling platform from multiple models. The purpose of the SCNG is to establish a global network for model inter-comparison on agricultural GHG emissions including the assessment of mitigation options, on the basis of collaborations, by coordinating common activities, facilitating data and protocols sharing and gathering international knowledge. It constitutes a support across GRA research groups in defining common objectives for improving models and methodologies related to soil carbon and nitrogen and to mitigation opportunities as they affect the production systems covered by the research groups of the GRA.

The first SCNG workshop was held in March 2011 in Orleans, France and focused on building a common set of objectives and work topics. A proof of concept of agricultural GHG model inter-comparison based on four models and two sites in Europe was presented and discussed during the second meeting of the group in Leuven, Belgium (2011). The work plan of the group was further elaborated and discussed during joint sessions with the CRG in Bari, Italy (2012) and with the LRG in Dublin, Ireland (2013).

The workplan included tasks of stock take of models and of site data, defining data needs for model benchmarking and inter-comparison, development of protocols for model inter-comparison, and identification with stakeholders of mitigation options to be evaluated by models. Moreover, capacity building and training steps for modeling and eddy flux measurements of GHGs were planned. The work plan was further discussed during side events at scientific conferences in 2012 and 2013 (in Sydney, New-York and Porto Alegre).

In March 2014, the SCNG organized in Paris a workshop attended by 80 participants from 24 countries to test 17 crop and pasture simulation models based on data from six sites covering four continents. Interactions with climate change are also being considered for inclusion in the work plan in close collaboration with AgMIP (www.agmip.org), MAGGnet, and the Australian National Agricultural Nitrous Oxide Research Program (NANORP).

Context and expected outputs: Crops and grasslands contribute to the biosphere-atmosphere exchange of radiatively active trace gases, with fluxes intimately linked to management practices. Of the three GHGs that are exchanged by agricultural lands, CO_2 is exchanged with the soil and vegetation, N_2O is emitted by soils, and CH_4 is emitted by livestock when grazing and from wetland soils. The magnitude of these GHG exchanges with the atmosphere vary according to several factors: climate, soil, vegetation, management, and global environment. Moreover, horizontal transfers of organic carbon to or from arable crops and grasslands may occur as a result of harvesting grass as silage or hay and from farm manure applications (Soussana et al., 2004). Additional carbon fluxes, which are minor in most cases, include leaching of dissolved organic carbon, carbon emissions during burning, carbon transfer by erosion, and volatile organic carbon emissions (Soussana et al., 2009). Management choices to reduce agricultural GHG emissions involve important trade-offs, e.g., preserving pastures and adapting their management to improve soil carbon sequestration may actually increase N_2O and CH_4 emissions. No-till may increase soil organic carbon stocks in arable cropping systems, but may potentially increase N_2O emissions from cultivated soils (Gregorich et al., 2005). As agricultural management is one of the key drivers of GHG emissions and removals, there is potential to reduce net GHG flux, expressed in CO_2 equivalents. However, this potential is highly variable at the field, animal, and farm scales since it varies widely with biological and ecological processes that are in part controlled by land use and land management (Soussana and Lemaire, 2014).

Tier 1 and Tier 2 GHG inventory methods provided by IPCC guidelines (IPCC, 1996; 2006) for the agriculture sector provide a strong and well-established methodology for national inventories, but these guidelines were not designed to calculate GHG emissions and removals at the field and animal scales. Therefore, only part of the GHG mitigation potential can be captured at these scales by Tier 1 and Tier 2 approaches. For instance, changes in soil organic carbon stocks induced by land management of arable and grassland systems (Kutsch et al., 2010; Soussana et al., 2010) are not well captured by IPCC guidelines. Moreover, interactions with soil and climate conditions that may lead to inter-annual variations in N_2O emissions and soil CO_2 fluxes are not accounted for by IPCC Tier 1 and 2 methodologies.

Indirect GHG emissions generated by farm activity through the use of farm inputs (e.g., fertilisers, feed, pesticides) are not accounted by the agriculture sector, but are covered by other sectors such as industry (e.g., for the synthesis and packaging of inorganic N fertilisers and of organic pesticides) and transport (e.g., transport of fertilisers and feed). Although the sectorial approach used by IPCC is appropriate for national and regional GHG inventories, it does not reflect emissions generated directly or indirectly by marketed products. Lifecycle analyses include indirect emissions generated by farm inputs and pre-chain activities (FAO, 2006; Weiss et al., 2012), but do not account for changes in carbon and nitrogen cycles triggered by detailed agricultural practices and by soil and climate variability.

In this context, the SCNG explores the potential of simulation models to predict the GHG balance of arable crops and grasslands as affected by agricultural practices. A stock-take has shown that more than 30 published models have been reported by member countries. However, these include various model categories, ranging from process-oriented models that require a large number of parameter values to simpler models that tend to use statistical rather than process-based inference. There are few generic models with the ability to simulate all GHG emissions and removals from both arable and grassland systems (e.g., DNDC, DayCent). Some models cover only soils (e.g., RothC, STICS) and others specialize either in grassland systems (e.g., PaSim) or in arable crops (e.g., CERES). Hence, delineating the validity domain of models and their prediction uncertainties, through inter-comparison and benchmarking over contrasted agricultural systems has been one of the first tasks by the SCNG. Recently, Asseng et al. (2013) showed that the impacts of climate change on wheat yields can best be projected by using the mean value of an ensemble of crop simulation models rather than by individual models. Further testing this hypothesis for the emissions and removals of GHGs at field scale is required.

Key outputs of the SCNG for 2014–2015 are:

- Consolidated list by region and by agricultural system of key mitigation options, agreed by stakeholders, to be simulated by crop and grassland models;
- Description of the validity domain of the main arable crop and grassland models for simulation of GHG emissions and removals and mitigation options;
- Inter-comparison and benchmarking of crop and grassland models and analysis of projection uncertainties for both individual models and for an ensemble of models.

Model inter-comparison and benchmarking activity: The first key activity of the SCNG is model inter-comparison, dedicated to quantifying GHG emissions and evaluating agricultural mitigation options. To this end, a stock-take identifying soil relevant carbon and nitrogen models and datasets available for comparison and collaboration was required. A preliminary exercise was used to launch the modeling activity through a pilot blind test with shared datasets (3 grasslands and 3 wheat crop systems) spread over 5 countries (France, India, Canada, New Zealand and USA). Various models and versions contributed to the exercise (Agro-C, APSIM, APSIM-GRAZPLAN, CERES-EGC, Daily Daycent, Daycent, DNDC95, DSSAT, EPIC, INFOCROP, Landscape DNDC, LPJG, PaSim, Roth C and STICS), resulting in 17 contributing models from 10 countries. Both sites and models were anonymous and observed vs simulated data were compared for each experimental site. This procedure helped to define main issues to improve modeling protocols and to extend the exercise with additional sites through a further blind test at a larger scale. Further steps are to calibrate models and to integrate mitigation options in modeling GHG emissions from grasslands and croplands. Information technology tools such as platforms to combine models are also being analyzed and networked through the SCNG.

Inventories and measurement cross-cutting group

The inventories and measurements cross-cutting group (IMG) is chaired by Canada and the Netherlands. The first meeting was held in November 2011 in Ottawa and subsequent meetings were held in Accra (2012) and Edinburgh (2013). The group aims to facilitate collaborative work to improve estimation and mitigation of GHG emissions in multiple segments of the agricultural production system.

A core activity of the IMG is developing an understanding of GHG emission intensity in agricultural systems, which is fundamental to fulfilling the goal of the Alliance. An important use of emission intensities is to understand relative GHG emission contribution of inputs, processes, and wastes to provide a product. A target of reducing GHG emission intensity is consistent with the GRA objectives of producing food with lower overall emissions than would have occurred without reduction in emission intensity. It has to be noted that with the reduction of emission intensity the total GHG emissions may increase, and this increase would be reflected in national inventories.

According to the overarching topic of GHG intensity, work is organized into: (i) remote sensing technologies to improve agricultural activity data; (ii) best practice guidelines for determining soil organic carbon (SOC); (iii) evaluating the economic value of GHG mitigation options; and (iv) developing decision frameworks for evaluating farming systems. Most work areas were initiated in 2013.

Remote sensing is expected to be a valuable tool in developing GHG inventories at regional scales. This is clearly reflected in land use change activity data, but remote sensing has the potential to be used complementary to more conventional farm level surveys. This work area will focus on remote sensing to obtain agricultural activity data. Agricultural activities such as crop rotation, irrigation management, and crop quality determine the two main components of GHG intensity, namely yield and GHG emissions. Work will start with a stocktake from participating countries by addressing: (a) spectral, spatial and temporal resolution, (b) geographic coverage and availability of images, (c) time period covered, and (d) costs and available technology to analyse data. A critical outcome of this work will be an overview of options of cost-effective remote sensing techniques to monitor and evaluate actual management at the field level.

Soil organic carbon (SOC) is an important attribute characterizing the status of soil, assessment of soil quality, and for quantification of soil CO_2 emissions and SOC storage. Guidelines for SOC determination are intended to assist scientists, senior technicians, and environmental professionals who will design and/or implement strategies for measuring SOC stocks and stock changes in agricultural lands. Guidelines aim to improve (a) comparability and consistency among different SOC measurement initiatives and (b) quality, accuracy and relevance of SOC measurements. Soil organic

C measurement for grasslands is an area where guidance would add scientific value to what is already more available on arable lands. Some of the topics that will be addressed are sampling strategies, comparison of inventory systems, sampling strategies over time, and spatial heterogeneity. The resulting guidance document will provide valuable information to countries that are considering a measurement scheme of SOC stocks, as well as a standard to countries that already have sampling schemes but that desire international uniformity to refine existing sampling schemes.

Work on the *economic valuation of GHG mitigation practices* is still to be defined. Costs at farm and regional levels are a critical element in the adoption of and support for GHG emission reduction measures—an area requiring attention by GRA-member countries.

Evaluating farming systems will be necessary to characterize a diversity of agricultural systems around the world. Climate change may reduce production levels with subsequent negative effects on GHG emission intensity. With adaptation and mitigation intrinsically linked at the farm level, we will focus on adaptation as part of a GHG emission intensity reduction strategy. Acknowledging the pivotal role of farming systems in decision-making is important when transitions and changes in farm management are needed to achieve food security or environmental goals. For both objectives, farm-level decisions and field management will determine the level of success. Farming systems provide a way to conceptualize agriculture and provide entry points for change towards more climate-friendly and less vulnerable agriculture. A workshop is planned to establish a farming systems network in the GRA. We will link to existing networks (e.g., International Farming Systems Association) and focus on how to address GHG emissions and resilience at the farm level. Creating a common understanding and approach to farming systems evaluation will be the main objective of this work area.

References

Asseng, S., Ewert, F., Rosenzweig, C., Jones, J.W., Hatfield, J.L., Ruane, A.C., Boote, K.J., Thorburn, P.J., Rötter, R.P., Cammarano, D., Brisson, N., Basso, B., Martre, P., Aggarwal, P.K., Angulo, C., Bertuzzi, P., Biernath, C., Challinor, A.J., Doltra, J., Gayler, S., Goldberg, R., Grant, R., Heng, L., Hooker, J., Hunt, L.A., Ingwersen, J., Izaurralde, R.C., Kersebaum, K.C., Müller, C., Naresh Kumar, S., Nendel, C., O'Leary, G., Olesen, J.E., Osborne, T.M., Palosuo, T., Priesack, E., Ripoche, D., Semenov, M.A., Shcherbak, I., Steduto, P., Stöckle, C., Stratonovitch, P., Streck, T., Supit, I., Tao, F., Travasso, M., Waha, K., Wallach, D., White, J.W., Williams, J.R. and Wolf, J. 2013. Uncertainty in simulating wheat yields under climate change. Nature Clim. Change 3(9): 827–832.

FAO. 2006. Livestock's Long Shadows: Environmental Issues and Options. FAO, Rome.

Gregorich, E.G., Rochette, P., VandenBygaart, A.J. and Angers, D.A. 2005. Greenhouse gas contributions of agricultural soils and potential mitigation practices in Eastern Canada. Soil Till. Res. 83(1): 53–72.

Intergovernmental Panel on Climate Change (IPCC). 1996. Revised Guidelines for National Greenhouse Gas Inventories. IPCC, Cambridge University Press, Cambridge.

IPCC. 2006. Good Practice Guidance on Land Use Change and Forestry in National Greenhouse Gas Inventories. IPCC, Institute for Global Environmental Strategies, Tokyo, Japan.

Kutsch, W.L., Aubinet, M., Buchmann, N., Smith, P., Osborne, B., Eugster, W., Wattenbach, M., Schrumpf, M., Schulze, E.D., Tomelleri, E., Ceschia, E., Bernhofer, C., Béziat, P., Carrara, A., Di Tommasi, P., Grünwald, T., Jones, M., Magliulo, V., Marloie, O., Moureaux, C., Olioso, A., Sanzi M.J., Saunders,

M., Søgaard, H. and Ziegler, W. 2010. The net biome production of full crop rotations in Europe. Agric. Ecosyst. Environ. 139(3): 336–345.

Minamikawa, K., Yagi, K., Tokida, T., Sander, B.O. and Wassmann, R. 2012. Appropriate frequency and time of day to measure methane emissions from an irrigated rice paddy in Japan using the manual closed chamber method. Greenhouse Gas Measurement Manage. 2: 118–128.

Soussana, J.F., Loiseau, P., Vuichard, N., Ceschia, E., Balesdent, J., Chevallier, T. and Arrouays, D. 2004. Carbon cycling and sequestration opportunities in temperate grasslands. Soil Use Manage. 20: 219–230.

Soussana, J.F., Tallec, T. and Blanfort, V. 2009. Mitigating the greenhouse gas balance of ruminant production systems through carbon sequestration in grasslands. Animal 4(03): 334–350.

Soussana, J.F., Graux, A.I. and Tubiello, F.N. 2010. Improving the use of modelling for projections of climate change impacts on crops and pastures. J. Exp. Botany 61(8): 2217–2228.

Soussana, J.F. and Lemaire, G. 2014. Coupling carbon and nitrogen cycles for environmentally sustainable intensification of grasslands and crop-livestock systems. Agric. Ecosyst. Environ. 190: 9–14.

Weiss, F. and Leip, A. 2012. Greenhouse gas emissions from the EU livestock sector: A life cycle assessment carried out with the CAPRI model. Agric. Ecosyst. Environ. 149: 124–134.

[1] USDA – Agricultural Research Service, 3218 Williams Hall, NCSU Campus Box 7619, Raleigh, NC 27695, USA* Alan Franzluebbers.
Email: alan.franzluebbers@ars.usda.gov

[2] Agriculture and Agri-Food Canada, Quebec, Canada.
Email: denis.angers@agr.gc.ca

[3] New Zealand Agricultural Greenhouse Gas Research Centre, Palmerston North, New Zealand.
Email: Harry.clark.nzagrc.org.nz

[4] INRA, Paris, France.
Email: fiona.ehrhardt@paris.inra.fr

[5] Queensland University of Technology, Brisbane, Australia.
Email: pr.grace@qut.edu.au

[6] Embrapa Headquarters – Executive-director of Research and Development, Brasília, Brazil.
Email: ladislau.martin@embrapa.br

[7] Agriculture and Agri-Food Canada, Swift Current, Saskatchewan, Canada.
Email: brian.mcconkey@agr.gc.ca

[8] Department of Agriculture Fisheries & Forestry, Australia.
Email: leann.palmer@daff.gov.au

[9] INRA, Reims, France.
Email: Sylvie.recous@reims.inra.fr

[10] Embrapa Soils, Rua Jardim Botânico, 1024. Rio de Janeiro/RJ.
Email: renato.rodrigues@embrapa.br

[11] INIA, Treinta y Tres, Uruguay.
Email: aroel@inia.org.uy

[12] Wageningen UR, Wageningen, The Netherlands.
Email: martin.scholten@wur.nl

[13] USDA – Agricultural Research Service, Beltsville, MD, USA.
Email: steven.shafer@ars.usda.gov

[14] Department of Climate Change and Energy Efficiency, Canberrra, Australia.
Email: bill.slattery@climatechange.gov.au

[15] INRA, Paris, France.
Email: jean-francois.soussana @paris.inra.fr

[16] Wageningen UR, Wageningen, The Netherlands.
Email: jan.verhagen@wur.nl

[17] National Institute for Agro-Environmental Sciences, Tsukuba, Japan.
Email: kyagi@affrc.go.jp

[18] INIA, Treinta y Tres, Uruguay.
Email: g.zorrilla@inia.gov.uy

* Corresponding author

4

Proposal for the Construction of a Greenhouse Gas Emissions Monitoring System for the ABC Plan

Sectoral Plan for Mitigation and Adaptation to Climate Change for the Consolidation of a Low Carbon Agriculture Economy

Giampaolo Queiroz Pellegrino,[1,a,*] *Renato de Aragão Ribeiro Rodrigues,*[2] *Aryeverton Fortes de Oliveira,*[1,b] *Eduardo Delgado Assad*[1,c] and *Luiz Adriano Maia Cordeiro*[3]

Historical Summary and the Current Status of the ABC Plan and Program

During the preparation for the 15th Conference of the Parties (COP15), in Copenhagen, Denmark, the Brazilian Government indicated the programs of climate change mitigation they proposed to adopt. The potential reduction in greenhouse gas emissions resulting from these action plans was estimated to be 35.1 to 38.9% of the projected emissions in Brazil until 2020. This voluntary commitment was made public through

Authors' affiliations given at the end of the chapter.

law number 12,187, of December 29, 2009, and plans were made in accordance to the National Policy on Climate Change, taking into account the specifics of each sector.

The Sectional Plan of Mitigation and Adaptation to the Climatic Changes for the Consolidation of an Economy of Low Carbon Emissions in Agriculture, also called the ABC Plan, is one of the Sectional Plans developed in accordance to Article 3 of the Decree 7,390/2010 and intends to organize the preparation of the action plans to be carried out in order to adopt sustainable production technologies selected to respond to the commitments the country has made towards the reduction of greenhouse gases emissions in livestock and agriculture. The ABC Plan is comprised of seven programs, six of which are related to the mitigation technologies and the last one with action plans of adaptation to the climatic changes. The plan takes effect nationwide and shall last from 2010 to 2020, with reviews and updates during this time.

The main goal of the plan is to "ensure the continuous improvement of the sustainable systems and practices in natural resource management, which will promote the reduction of greenhouse gas emissions as well as increase the fixation of atmospheric CO_2 in the vegetation and the soil on the different sectors of the Brazilian agriculture".

Its main commitments are strategies that involve the following technological maneuvers:

- Recover an area of 15 million hectares of pastures degraded by the use of adequate management and fertilization;
- Increase the employment of Crop-Livestock-Forest Integration systems, as well as Agroforestry Systems in four million hectares;
- Expand the use of the No-tillage System in eight million hectares;
- Biological Nitrogen Fixation: expand the usage of biological fixation in 5.5 million hectares;
- Promote reforestation actions in the country,[1] expanding the area of Planted Forests, currently destined to fiber, wood and pulp production in three million hectares, therefore going from six million to nine million hectares;
- Amplify the use of technologies for the treatment of 4.4 million cubic meters of animal waste so as to generate energy and produce organic compounds.

As far as the creation of finance, Resolution number 3,896 of the National Monetary Council of August 17, 2010, has established the Reduction in Greenhouse Gas Emissions in Agriculture Plan (ABC Plan) within the scope of the programs with resources from BNDES. The ABC Program intends to benefit industrial agriculture and, aside from investment funding, its purpose is to finance fixed and semi-fixed investments, destined to the recovery of degraded areas and pastures, to the implantation of integration systems, such as livestock-crop, forest-crop livestock-forest or crop-livestock-forest, as well as the implementation and maintenance of

[1] Take note that in this additional effort in agriculture, the Brazilian commitment concerning the Metallurgy Sector is not calculated.

commercial forests or those destined to the restoration of legal reserves or permanent preservation areas.

Several monitoring actions of the ABC Plan are already underway, and the great majority of these actions focus on monitoring the execution of the activities of the Plan itself. However, it is necessary to develop an integrated monitoring system that allows appropriate administration and continuous improvement of the ABC Plan and Program. Furthermore, it should be a system that is based in specific greenhouse gas emissions monitoring protocols that are developed, preferably, in accordance to the regulations implemented under the UNFCCC, focusing mainly on the post-2020 scenario.

This chapter intends to propose the structure and implementation of this monitoring system, relying on the administration and participation of several stakeholders in different areas, essential to the innovation and development of the Brazilian agricultural sector.

Implantation stages of the ABC plan integrated monitoring system

In order to recommend monitoring methodologies for the reduction of emissions by adopting the technologies proposed in the ABC Plan, this chapter presents two monitoring scenarios:

Emergency scenario (2014)

While the Ministry of Agriculture, Livestock and Supply does not assemble the integrated general monitoring system of the ABC Plan, which in our opinion should contemplate the elements and premises described in the operational scenario of actual information monitoring by the system, we recommend that the gas emission reduction due to the employment of the technologies proposed by the ABC Plan be roughly estimated during the year 2014, as an emergency transitional period.

In this scenario there is no incorporation of new research results or new emissions factors specific to the country (or the regions). Also, there is no satellite monitoring for the properties that adopted the ABC Plan technologies (independently of the use of resources of the ABC Program).

According to baseline surveys previously carried out by Embrapa and partners, we would consider only the extrapolation of the potential reduction in greenhouse gas reduction by the technologies in relation to the area occupied by these technologies with resources of the ABC Program. The area would be calculated from the information given by the farmers to the bank system. The calculation could be done by property, or to partially or totally monitor the properties that received the fund. It is strongly advised that there be a current stage of monitoring (during 2014), with the records of all properties that already have the fund. A new monitoring stage should be carried out at the end of each harvest-year, until 2020.

Monitoring per area involves almost all the Plan technologies (recovery of degraded pastures, Crop-Livestock-Forest Integration, Farming-Forest Systems, no-tillage system, nitrogen biologic fixation and planted forests). With an estimate of the area, we would be able to calculate the gas emissions by using standard emissions factors, associated to each technology.

Besides the use of the area to monitor the emergency scenario, adding other indicators, such as: baseline soil carbon stocks in the recovered pastures and the areas with Crop-Livestock-Forest Integration, Agroforestry Systems, no-tillage and planted forests; the number of vaccine dosages sold in the case of the nitrogen biologic fixation; and volume of processed gas, volume of methane used in the generation of electricity with the use of biogases, as well as tons of organic compounds generated for the waste treatment technology.

We also suggest adding to this monitoring methodology, the reduction of emissions promoted by projects of Clean Development Mechanisms (CDM) for the planted forests and waste management technologies.

The emissions reduction potential promoted by Clean Development Mechanisms projects already under implementation is considerable and must be taken into account by monitoring the sectoral plans. In the case of swine, which would correspond to the animal manure management technology, 77 projects in approval/validation stage had been registered by June 30, 2011, and together they reduce approximately 4.3 million tons of CO_2e per year. It should be stressed that the total goal for this technology under the ABC Plan is 6.9 million tons and that Brazil still has a great potential for the expansion of this activity and the implantation of the methodology of Clean Development Mechanisms in swine farms.

On the other hand, reforestation still has only three activities projects, but the potential in this sector is considerable. All three projects of reforestation currently in the approval/validation stage reduced more the four millions RCEs[2] up to December 2012.

Nevertheless, with the scenario described above, we will still have imprecision in the calculation of the emission mitigation estimates.

Definitive scenario (operation from 2015 onwards)

For the development of the definitive or operational scenario, an integrated monitoring system must be created for the ABC Plan. This system must be able to report the reduction of emissions and be verified in accordance to international regulations, which will allow the integration of the different elements of the emissions monitoring system attributable to the use of the ABC Plan technologies, implemented with the resources from the ABC Program. For that reason, the following elements will be needed:

- The monitoring system should be tied to the banking credit system and their respective generation, organization and supply of information, including the System of Rural Credit Operations and Proagro (SICOR), as well as additional data needed for the registry of fund proposals;
- Satellite monitoring of the implemented technology areas for projects financed by the ABC Program;

[2] Each RCE is equivalent to 1 tonne of CO_2e.

- Creation of a greenhouse gas emission database in Brazil, which shall be compatible with the current rules, especially with the International Consultation Analysis (ICA) and the inventory procedures of UNFCCC for the period after 2020;
- Development of emission factors specific to Brazil (with preference to the development of regional emission factors) for the ABC Plan technologies.

With this information, it will be possible to create a robust mechanism for monitoring emissions reductions, in addition to the adoption and efficiency of the Program. This system suits the international requirements that the action plans be measurable, reportable and verifiable (MRV in the jargon of the Framework Convention on Climatic Changes from the UN).

With this data organized, it will be possible to create a proposal for a web tool for emissions reductions, based on the database with information about the type of culture, soil, topography, climate, management (includes culture system, mechanization, fertilizer and agro toxics used, etc.) and emissions measured *in loco* and calculated.

This tool will allow the calculation of projects emissions reduction financed by the ABC Program, which will generate a monitoring system for emissions reduction with high reliability. This data can also subsidize the development of some of the state or national policy for the Payment of Environmental Services.

Description of the ABC plan integrated monitoring system (SiM-ABC)

The emissions reduction monitoring, which will be carried out by the ABC Plan's Multi-institutional Monitoring Laboratory during the operational scenario proposed in this chapter, can only be conceived, planned, projected and completed in all its aspects if there is an integrated general monitoring system for the ABC Plan. In this regard, we recommend following the ABC Plan Integrated Monitoring System (SiM-ABC), which has the essential elements to enable the Multi-institutional Monitoring Laboratory of the ABC Plan (LiM-ABC) to properly monitor emissions reduction, since it is only one of many elements and will not be able to fulfill its demands without the existence and proper functioning of the others.

Main purpose of the ABC plan integrated monitoring system (SiM-ABC)

The ABC Plan Integrated Monitoring System (SiM-ABC) focuses on promoting the integration of the efforts of several stakeholders, including:

- Estimates and monitoring of the emissions reduction and the increase of carbon stock in the production systems that received funds under the ABC Program;
- Monitoring of proper use of this credit, by following up with actions employed according to what has been established in the ABC Plan; monitoring its adoption by monitoring the actions implemented as provided in the ABC Plan;

- Monitoring of the economic and technical efficiency production as well as the sustainability of the undertakings financed by the ABC Program.

Each of these SiM-ABC components or elements will have different institutional responsibilities, and the general SiM-ABC administration will be done by the National Executive Commission, lead by the Ministry of Agriculture, Livestock and Food Supply (MAPA—acronym in Portuguese) and the Ministry of Agrarian Development (MDA—acronym in Portuguese). On the other hand, the mitigation and adaptation actions to climate change scheduled in the ABC Plan to be carried out by 2020 will be under the joint responsibility of MAPA and MDA. The emissions reduction monitoring and the carbon stock monitoring in the production systems that received the finance under the ABC Program will be established by the LiM-ABC, and the supervision of the implementation and the technical and economic efficiency in the production and sustainability of funded projects by ABC Program will be the responsibility of MAPA, MDA and the National Banking System (Fig. 4.1). Despite these specific institutional responsibilities, each components of SiM-ABC will involve the collaboration and co-responsibility of all these institutions.

In this regard, SiM-ABC becomes an essential instrument in the continuous improvement and the administration of the action plans and the policies of the Brazilian Government regarding climatic changes and its interactions with the agricultural sector, and the three main highlights (shown in Fig. 4.2) are based on the fact that the ABC Program can only succeed if the improvement in economic efficiency attributable to the finance granted under the program is proven. In other words, we believe that, as concerned as the farmer may be with environmental matters and with the impact of climatic change over their production, they must focus on the economic viability of their enterprise. On the other hand, it is also understood that the opportunity offered by the program must create mechanisms in order to avoid misuse of the resources, especially by opportunists who may access the program without the commitment to follow through with the requirements of the contract.

With this in mind, the implementation of the action plans financed by the ABC Program may be enough to guarantee compliance and performance of the voluntary commitment proposed by the Brazilian Government, after COP15, in Copenhagen, 2009, and the Sector Plan of Climatic Changes for Agriculture if it is demonstrated to farmers that, besides the environmental benefits of adopting the fund, its economic co-benefits are unquestionable, either by increase in their revenue, or by reducing their losses which can already be realized or planned for the future.

This monitoring proposal also comprises of multiple benefits within the Program, which must be reached and valued along their development. The productive intensification subsequent to the implementation of the ABC Plan strengthens the offer of food without opening native vegetation, a kind of benefit that may favor environmental services provided by the remaining vegetation, improvement in the rural landscape and the conservation of resources for future generations, environmental values that must be observed and used for the promotion of the Brazilian agriculture in future opportunities. It will also make it feasible to officially distinguish the products and productive chains involved in the national and the international markets.

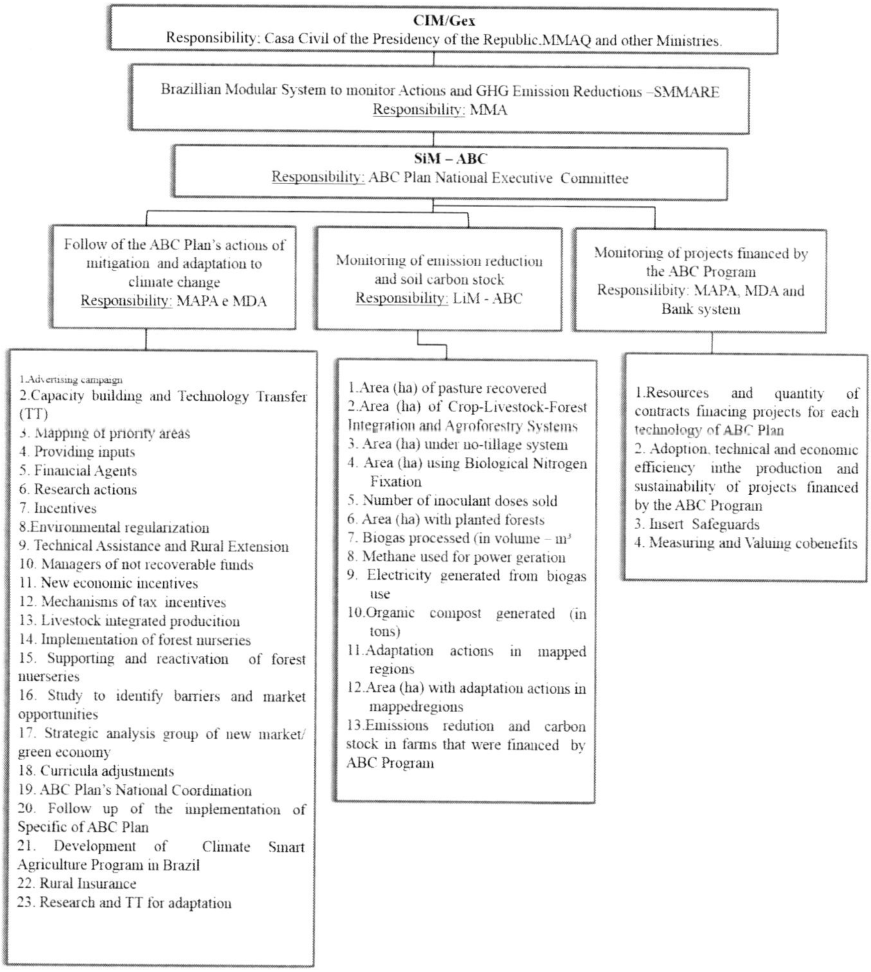

Figure 4.1. Administration chart and monitoring indicators according to the ABC Plan.

The necessity of the initial survey and the periodical data monitoring focused on these three main points:

- The adoption of the ABC Program—Information on the Project Proposals to be funded, detailing its performance and filled out in digital format, containing, in specific fields, the location, area, fund and registration sketch developed on map systems and online satellite imaging, such as Google Maps.
- Efficiency improvement and the sustainability of the credit grantees—information on the economic condition of the enterprises to be financed and monitoring its evolution, according to the procedures already adopted by the banking system for the fund evaluation under the ABC Program, among others. Social and environmental information associated to the farmers and the location of the implementation of the Plan complement this efficiency analysis; and

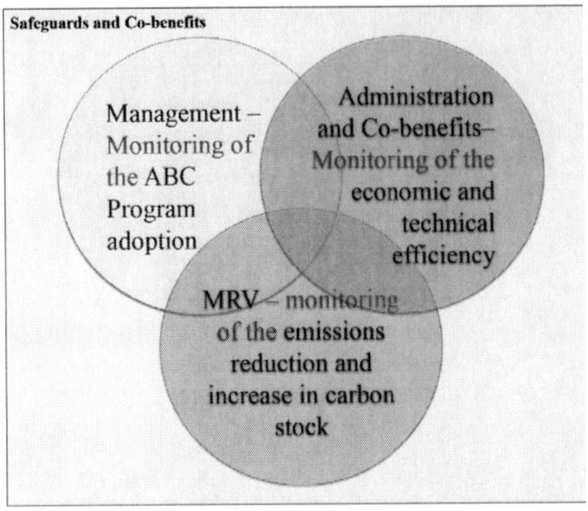

Figure 4.2. Chart representing the three main focus points of monitoring, its intersections, its contribution in the survey of data for governance and analysis of the benefits to the Sustainable Development.

- Emission mitigation in agriculture and the fulfillment of the goals under the ABC Plan—Information about the greenhouse gases emissions[3] and/or the initial carbon stock in the area and the production system of the financed project, as well as the advancement caused by the fund obtained.

Survey of Information on the Adoption of the ABC Program

For this survey, information about the proposed project to be financed will be organized, with details of their implementation and filled in a digital format, containing, in specific fields, complete address (via, county, state, zip code, etc.), the central point geographical coordinates, area, funded and registration sketch developed over maps and satellite images in computerized platform with access via Internet system.

During the emergency scenario described above, this information would be enough for the Multi-institutional Laboratory, responsible for the ABC Monitoring, to prepare estimates of the efficiency of the program, also in terms of emissions reduction in agriculture.

In the SiM-ABC operational scenario, the technician designing the project would obtain information, record it and send the files containing the specifications of their proposal and project by using an efficient and specific software, to be developed with the assistance of Embrapa Agricultural Informatics or another outsourced company. This software would serve as the initial interface between the proponent and the banking system, since it would allow the inclusion of the information already required by the banks for finance for the ABC Program, by SICOR, the Rural Credit Transaction System and Proagro, in addition to other information that may be needed for the proper evaluation of the implementation and efficiency of the program.

[3] As the emission data are very difficult to obtain, in some cases, only the initial carbon stock data will be available.

The information would be stored in a central database that will allow the integration of the information in the system for the evaluation of the effectiveness of the ABC Program, primarily but not exclusively for the emissions reduction monitoring due to the adoption of technologies financed by the ABC Program. As long as it is viable and it does not threaten the security of the system, as well as being authorized by SICOR management, this system would already feed SICOR with information required by them, allowing one single entry of information by the banking system.

The official survey of the current status of the implementation project area, and possibly monitoring of the evolution of new practices financed in that area will be based on a set of high resolution images every two years for the entire national territory, such as the images 'Rapid Eye' already obtained by the Ministry of the Environment (MMA—acronym in Portuguese) during the years 2010/2011. Also, lower resolution images shall be widely used, which will allow the supervision of the evolution of the culture, and support the analysis of the implementation of the practices financed in these areas.

After the projects implementation, technicians responsible for the projects, researchers and assessors within the banking system can also carry out the monitoring action plans periodically. There will be a need to develop an interface and an information management system that will offer support to the iconographic georeferencing registry of the practices implemented in the financed areas, input of the information on soil analysis, especially about the carbon stock in these locations and the information about the status of the financed projects.

Additional information shall be obtained by questionnaires, to be answered in the field at different stages. These questions would focus mainly on surveying the knowledge about the program, the reasons for adopting or not, and perceived benefits and difficulties for the funded program. Trained technicians would carry it out during field visitations with national and regional circuits, using interviews and/or displays with technicians and specialists, in what could be called 'ABC Rally', similar to the 'Harvest Rally' (Rali das Safras).

As soon as possible, specific studies should also be carried out for each individual fund so as to map the actual current conditions and prioritizing the regions, groups and associations, specific sectors of the agriculture and other adequate forms of stratification that must be the target of motivational action plans and breaking the barriers to the program implementation. These studies must combine both specific data survey and the information on the efficiency in the emission reduction and economics, among others, for the establishment of appropriate strategies for the increase of efficiency of the Program and, consequently, of the ABC Plan.

Information Survey on the Borrowers Efficiency and Sustainability Improvement

For the rural sector credit analysis, the banking system already requires and registers information regarding the initial economic situation of the target enterprises. In the event the financing is granted, the banking system reserves the right to collect information for periodic evaluations for their own purposes, such as the supervision of the effectiveness of the employment of the resources received and risk assessment for future insurance claims.

For this reason the registration information and the technical reports become important elements of rural credit management and, consequently, for the ABC Plan. The three biggest sets of data described as the focus of the ABC must be accessible and also be supplied by the banking system, for it is an important part of the ABC Plan management and necessary for its continued improvement. This information can be incorporated to the SiM-ABC to be analyzed collectively or even separately for each financed project, always keeping total confidentiality, so as to keep from identifying the specific project to which it refers. In general, beside from the financial data itself, this involves data about the products financed, such as fertilizer, machinery, improvements, irrigation systems, animals and others.

As mentioned before, the analysis of this information always results in benefits to the efficiency of the ABC Program. From the farmer's point of view, if it is proven that there is economic growth resulting from the financed projects and the difference in the product or the productive chain in the national and international markets, the program will be able to show that not only does it benefit the environment, but also the farmer, thus becoming an interesting option. Its economic co-benefits are undisputed, be it by diversifying or increasing income, by the decrease in losses, which may be noticeable now or in the future.

From the perspective of the ABC Plan and Program management, if a decrease or absence of effects on the economic growth of the farmers as a result of the financed projects is noticed, they will be able to correct, redirect or even recreate strategies and funds to guarantee compliance with the voluntary commitment proposed by the Brazilian Government, after COP15, in 2009 and in the Sectoral Plan of Climatic Changes on Agriculture, the ABC Plan. Moreover, it is an important instrument to make a case in favor of the efficiency of the national productive systems in relation to potential commercial litigation within the World Trade Organization, especially about matters relating to the 'Green Box' and commercial rules specific to the agricultural sector and sustainability.

Therefore the survey and monitoring of the economic information of the financed farmers are not only beneficial, but essential. This is also related to the creation of a specific system that may support the banking system in collecting information on the credit applications and the approved loans, which can be replicated and integrated at the same time it is able to maintain confidentiality and privacy of the grantees.

Associated with this economic aspect of the financing program, as well as the guarantee the program will be adopted and its goals reached, is the argument made by FGV (2013) about the different roles that could be played by Banco do Brasil (Brazil Bank), mainly for its widespread reach, technical team and proximity to the farmer, in addition to The Brazilian Development Bank (BNDES), for its capacity to finance projects that are inclusive, larger and with more structure.

The document debates: "One of the hypotheses for the low initial rates of implementation of the ABC is the fact that the BNDES operations have not been adapted to the structure and demands of the program.... The low number of applicants must be evaluated in order to improve the resource management and increase the implementation rates... due to the low amounts per project that were loaned... the operational costs for the bank ought to be too high and have low coverage. A deeper analysis of these facts must be a top priority, aside from short-term solutions that will increase the effectiveness of the ABC Plan".

This matter is also related to the laboratory infrastructure implementation for soil carbon analysis, as well as other elements of SiM-ABC, described in this document, where BNDES probably would have a greater efficiency in the distribution of the allocated resources. However, with the recent 'BNDES card' system and its operation connected to several banking institutions which, together, overshadow the widespread coverage of the Banco do Brasil agencies, it would be enough to adapt their rules and regulations to make BNDES more efficient than Banco do Brasil, even when it comes to providing credit to small and medium sized rural businesses.

As for the maintenance or improvement of the efficiency and sustainability of the financed projects, aside from the economic information, data on the environmental and social variables associated to the farmers and localities financed should also be collected, so that they may be analyzed and safeguard systems may be established for special situations in the interest of preservation in conformity to the laws and regulations, which must be clearly stated in the analysis of proposals for technologies and productive systems during the implementation of the Plan. The information on multiple benefits generated by the productive systems supported by the program must also be carefully registered.

Survey Information on the Emissions Reduction from Agriculture

The goal of the ABC Plan is the emissions reduction in agriculture, fulfilling the voluntary commitments presented by the Brazilian government, such as Nationally Appropriate Mitigation Actions (NAMAs) after COP15 and the other complements established in the plan. At the heart of this matter are the initial survey and the monitoring of indicators that may express this reduction and allow international acknowledgment of the data presented and, consequently, the fulfillment of the established voluntary commitments.

More than that, the tendency that is forecast with the development of the international negotiations is that Brazil and other Parties not included in Annex I may have additional commitments, including transparency, established in a new agreement that will enter into effect beginning in 2020.

The indicator for reducing emissions from agriculture is based on the estimative of the area covered by ABC Plan's technology, the carbon stock or, by the use of the definition of emissions factors, the initial greenhouse gas emissions in the area and production system of the financed project, expressed in the form of CO_2e (equivalent carbon dioxide), aside from its evolution owed to the acquisition of the fund.

However, other indirect indicators, that may be accepted internationally and monitored for the financed projects, are obtained in accordance to its characteristics and peculiarities. For the projects already financed, in exceptional cases that respect the safeguards previously established and allow the control of opportunism though the use of technical and economic monitoring, a technical-scientific committee may approve alternative indicators of the efficiency in emissions reduction, other than the samples and quantification of carbon stock or the originally defined indicator, as long as they pass through the same process proposed for the inclusion of new funds under the ABC Plan, described in a specific item in this document, being exempted from the deadlines needed for the revision of this plan.

From the international agenda point of view, it is necessary to emphasize the need to ensure the mitigation efforts historically reached by Brazil before 2020, which must be properly documented and monitored, both from the qualitative point of view regarding the environmental and social benefits, as well as quantitative with regards to the fraction of carbon or equivalent carbon effectively reduced.

Adequate monitoring of these actions will be a *sine qua non* condition to effectively ballast Brazilian arguments that seek the treatment of these efforts in an appropriate manner in the list of historical responsibilities that all countries must take upon themselves after 2020, in accordance with the principles and provisions of the UNFCCC, in particular the principle of 'Common but Differentiated Responsibilities'.

The emissions reduction monitoring reached by activities related to the ABC Program must, just as all the other NAMAs, be published through an instrument called BUR (Biennial Update Report) and a process of verification called ICA (International Consultation and Analysis). For this reason, it is highly recommended to integrate the Clean Development Mechanisms and the emissions reduction promoted by its projects, directly or indirectly related to the agriculture and its productive chain as an additional tool to the ABC Plan.

This information survey on agriculture emissions reduction due to the fund given by the ABC Program is the focal point of the ABC Plan Multi-institutional Monitoring Laboratory, but it will only be possible with the structuring of the SiM-ABC with their essential elements, described as follows.

Essential Elements to the Operation of SiM-ABC

For the success of the information survey and monitoring of the three pillars proposed by SiM-ABC and for its continued improvement, some elements of the system have a primary role in its operation. They are:

- The process for the inclusion of new funds;
- Regional training of farmers, consultants, bank representatives and agrarian sciences technicians regarding the plan and its regional adaptation within the principles accepted internationally and with participative structured methodology that allows to improve the discussion involving the main agents and guarantee the maintenance of the focal point and direct concrete results on the adaptation of the program to the reality of local systems;
- The laboratorial infrastructure for the analysis of the total soil carbon, with a connected system of accreditation of technicians and specialists, trained to collect soil with adequate methodology and frequency, previously established by SiM-ABC, and whose costs would be included in the amount financed for the project; and

- The ABC Program Multi-institutional Monitoring Laboratory (LiM-ABC), that are already established monitoring activities and are detailed in a specific item of this document.

Figure 4.3 presents these elements of SiM-ABC, as well as the many stakeholders involved and the flow of information for effective monitoring of the ABC Plan and the emission reduction in this sector. The numbered arrows show the sequence expected in the information flow, from the training of the several agents and the elaboration of proposals to be financed, to its concentration and utilization by LiM-ABC and its advertising, as well as the inclusion of the fund that will go through this same flow.

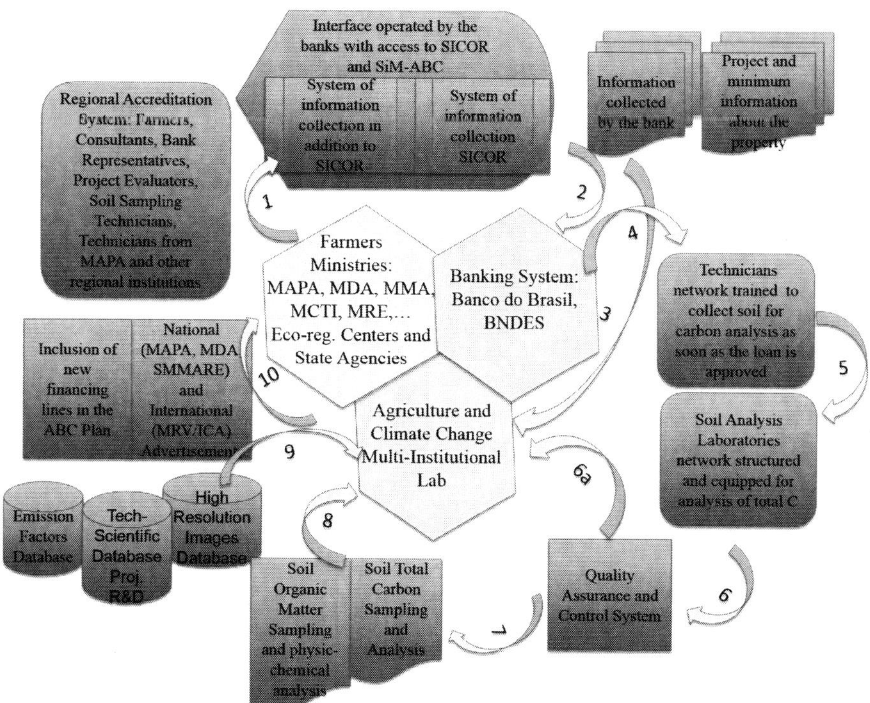

Figure 4.3. Elements, agents and flow of information of the Integrated System of Monitoring of the ABC Plan (SiM-ABC). The information flow between the elements of SiM-ABC occurs in the same sequence as the arrows, numbered from 1 to 10.

Proceedings for the Inclusion of New Finance

The ABC Plan periodical revision has a previously established regularity and, it is only within these that it would be possible to include new funds within the ABC Program. For this inclusion, it is necessary to define a process to ensure a detailed analysis of the possible implications for the ABC Plan success and the record of lessons learned to support decision-making at the domestic level as well as the positioning of Brazil under the UNFCCC in the period post 2020.

The proceedings proposed here, as an essential element of the SiM-ABC, and the following steps are expected to be followed for the inclusion of the lines of credit within the ABC Plan:

1. The proponent interested in the new fund shall prepare a document that proves scientifically, beyond any doubt, the real emissions reduction of the production system it refers to, following orientations and criteria defined under the coordination of MAPA and LiM-ABC and approved by the ABC Plan National Executive Commission, and fill out the form that has the specific mandatory items, which will include, necessarily, the quantification methodology in the emissions reduction, the parameter and indexes that shall be monitored and the appropriate collection methods, among other information yet to be identified. If there is public interest or representation of small family or underprivileged groups, MAPA itself will be able to assume the role of the applicant.
2. The proposal contained in this document will be forwarded to the National Executive Commission, MAPA and LiM-ABC, for analysis of the requisites and the selection of consultants and technical-scientific institutions specialized on the subject in question.
3. The consultants and the institutions selected will perform an impartial analysis of the relevance or not of the proposal and the convenience of its inclusion in the ABC Plan, always keeping in mind the national goals of this plan, and will write an opinion that synthesizes the proposal's main points, highlights the pros and cons and recommends or not its approval.
4. The National Executive Commission, MAPA and LiM-ABC will receive and analyze the opinion and give technical assistance to the National Executive Commission, which will deliberate on the matter and decide whether it should be submitted to the Federal Government for evaluation.

The National Executive Commission, MAPA and LiM-ABC will be responsible for organizing and establishing the orientation and application forms for the new funds inclusion. Embrapa and other technical-scientific institutions related to the Federal Government Ministries that may be involved, must reserve the role of application evaluators, defined in steps 2 and 3 of this process. Therefore, they should not take on the role of applicant or assistant to applicants, as described in step 1.

ABC Program Training and Regional Adaptation

The success of the ABC Plan is closely related to the wide coverage of ABC Program financing. With this in mind, the fund and/or its indicator adequacy and monitoring methods to the regional reality of the farmers and production systems are essential. This can be done by introducing new funds at the moment of Plan revision or the adaptation of the indicators or monitoring methods of funds already approved, which can be done at any time, but must follow the same process and criteria defined for the introduction of a new fund, as proposed in this document.

The ABC Plan regional training and its regional adaptation must happen within the internationally accepted principles and established by the Plan and must be based on a participative methodology previously established, applied and facilitated by a qualified technical team, in order to make direct and efficient discussion involving the main stakeholders and ensure the maintenance of focus and achievement of concrete results about the program adaptation to the reality of the local systems. For this purpose, we suggest that MAPA coordinate these actions involving its state offices and other institutions of several different areas, such as Embrapa's eco-regional offices, State Organizations of Agriculture and Livestock, extension agents, banking system, cooperatives and associations of farmers, to name a few. The discussion of pilot actions for the livestock systems in the Pantanal region and in Minas Gerais have already been conceived by Embrapa's researchers, proposing this regional adaptation and stakeholders integration, noticeably MAPA, the farmers and the banking system.

Although all levels must have knowledge about the history, goals, actions of the ABC Plan and ABC Program funded their lines and upgrades, training should also be given in a specific way according to the different roles to be played within SiM-ABC.

- Farmer proponents must have knowledge about the possible adaptations of their production system by means of good practices stimulated by the ABC Plan, the environmental, social and financial advantages promoted by its adoption, the economic advantages of the low interest rates offered by the ABC Program, the forms of monitoring and the necessity to collect periodic samples by accredited technicians, and its obligations and responsibilities, as well as penalties associated with them;
- Project developers consultants must have extensive knowledge about the process of submission and evaluation of project proposals, the principles and information required to be included in the project, filling out forms, the methods and indexes to be monitored for each fund, their obligations and technical responsibilities and the commitments taken on by the proponents;
- Banking staff project evaluators and information registration must be knowledgeable about the submission and evaluation process, the principles and information required on the project, the criteria of evaluation required by the ABC Program, filling out forms, indexes to be monitored and, mainly, about the information registering digital system, whether required by SICOR or by SiM-ABC;
- Accredited technicians to collect soil samples and other monitoring elements must have detailed knowledge about the methods, indexes to be monitored and sampling methodologies required for each fund, the principles and information required on the project and reports, filling out forms and their obligations and technical responsibilities within the auditioning criteria defined by the Plan and by the finance system;
- MAPA technicians and other institutions involved in the Program training and regional adaptation, aside from knowledge about the elements and stages of SiM-ABC, must know the techniques in training and methodology of efficient and objective ways to carry out workshops in order to obtain the results expected.

Laboratory Infrastructure Improvement and Soil Sampling

Taking into account that the burden of carbon analysis must not be placed on the farmer and recognizing the importance of this measure for the ABC Plan monitoring we add the importance of international recognition in order to reach the goals of the ABC Plan and the commitment by the Brazilian Government to the United Nations Framework Convention on Climate Change (UNFCCC). Considering these arguments, we highlight the importance of investments for the improvement and expansion of laboratorial infrastructure to improve its distribution and access throughout all national territory, aside from increasing the capacity of soil total carbon analysis.

It is seen that the Brazilian Central-West, Northeast and North regions lack these services of analysis, which compromises the adoption of a technological package that is more efficient and productive for Brazilian agriculture. More importantly, there is great difficulty to carry out carbon analysis which is almost nonexistent currently.

It is important for this to be done because, in connecting these actions to the ABC Plan, there will certainly be a more rational consumption of mineral and organomineral fertilizers in Brazil, thus reducing greenhouse gas emissions. At the same time, the correct use of fertilizer recommendations must result, over time, in the increase of soil organic carbon and the increase in the food production.

Aided by the ABC's Observatory Report FGV (2013), the proposal, in this case, is to have BNDES finance a national study regarding the real demand in order to understand the ABC monitoring, as well as other studies that may need this type of analysis, and focus their collaboration to ABC in this expansion of the laboratorial infrastructure and other structural projects, which are projects of larger amounts, connected to the improvement of the Brazilian agricultural logistics and more, according to the profile and expertise of that financing institution. On the other hand, with the 'BNDES Card' this bank, together with Banco do Brasil, would increase its role in the ABC Program as a project lender, even the smaller ones, given the increased coverage to work closer with the farmer and by smaller loans.

Connected to this demand for soil total carbon analysis infrastructure, it is necessary to ensure that the soil samples for ABC monitoring be carried out at an adequate location, according to the methodological protocol previously established, and by accredited trained technicians. Similar to the inspection of initial conditions for any credit or insurance granted by the government or the private sector, this activity would guarantee the samples quality and avoid fraud, given that, without it, there could be no certainty about the authenticity of the analyses presented by the farmer. Furthermore, it would make it possible to obtain and check information about the area that may receive the finance. This sampling and analysis would be repeated with the regularity determined by LiM-ABC, and its costs, which would include the physic-chemical and organic matter analysis established in the new text proposed by GT-ABC, and be included in the amount financed without becoming a burden to the farmer prior to the acceptance of the proposal.

Multi-Institutional Laboratory for the Monitoring of the ABC Program (LiM-ABC)

Given its primary role within the elements of SiM-ABC, and the necessity of improved knowledge of its activities and methodologies, it is described in further detail, highlighting that some activities and proposals of technical-scientific projects have already been in progress.

The Multi-Institutional Laboratory for the ABC Plan Monitoring (LiM-ABC)

The Climatic Changes and Agriculture Multi-Institutional Laboratory which will be the institution responsible for the ABC Plan monitoring especially regarding greenhouse gas emissions reduction in agriculture, has finished the stage of its physical structure construction, and it is expected to begin activities in 2014. Its detailed operations project, containing the aspects "description of management of the project at administrative, technical and budgetary levels; methodology employed in the collection and processing of data; technical and administrative team; budget specifications; mechanisms of data processing and accessibility; among other aspects considered relevant for the best description and formalization of the project" will only be finished with the structuring of SiM-ABC and the creation of the LiM-ABC Management Committee, comprised of representatives of the institutions that will integrate the Multi-institutional Monitoring Laboratory of the ABC Plan.

On this note, given its primary role within the elements of SiM-ABC, and the necessity of improved knowledge of its activities and methodologies, it will be described in further detail, highlighting that some activities and proposals of technical-scientific projects are currently ongoing.

Details of the LiM-ABC role and methodology

With the purpose of monitoring soil carbon stock in iCLF areas and degraded pastures, more than 900 soil samples were collected at the depths of 0–5, 5–10, 10–20 and 20–30 cm, in properties that use the integration production system Crop-Livestock-Forest and no-tillage system. These samples are under analysis in several research institutions and the first results were already released in 2013. Such analyses, supported by a solid network of Technical Reference Units, if used with adequate regularity throughout time and spatial distribution, will serve as a basis of reference for the comparison of the data collected in the areas of the financed project and will be essential in the LiM-ABC emergency scenario, described above.

Although the LiM-ABC activities have a highly operational character in relation to the emissions monitoring in agriculture, the technical-scientific support is essential for the methodologies and indicators used. In this regard, Embrapa, in its strategic research redirection, since 2007, has been working in the creation of a research framework of climatic changes that is structured in its Research Projects and Climatic Changes Portfolio, which currently has over 60 projects.

Among these projects, there are three large projects specifically focused on the balance of carbon in livestock, crop and forest systems, and also integrated systems where we determine the emissions and carbon storage on the productive systems and through which we wish to give the main scientific support to the activities in the national inventory of gases and the methodological definitions for the operational emission reduction monitoring in agriculture to be carried out by LiM-ABC.

Moreover, many other projects financed by the Embrapa Management System and other sources have been carried out and proposed, so aside from what is available in the national and international literature, we may obtain results that are specifically applicable to the national agriculture systems.

Aiming to determine the specific emissions factors for the integrated Crop-Livestock-Forest system, Embrapa Agrosilvopastoral (Sinop, MT), has begun a project financed by the Ministry of Environment and Embrapa itself, where we acquired an automated system of chambers for greenhouse gases collection and analysis. This equipment, manufactured by Queensland University of Technology, is the first of its kind in Brazil, and is now part of the international 'N_2O Network', together with Australia, United States, India and Chile. Also at Embrapa, at the beginning of 2014, a project of greenhouse gas emissions measurement was begun, testing several different levels of fertilization and the use of inoculators for corn crop. In a similar initiative, Embrapa Rice and Beans, in Goiás, is already operating an automated system for greenhouse gas measurement, that operates based on micro-meteorological processes. The use of automated systems is extremely important to obtain emissions factors.

Another project funded by the Ministry of Environment and Embrapa, to be carried out by Embrapa Agrosilvopastoral and Embrapa Southeast Livestock (São Carlos, SP) will study the emissions of enteric methane in integrated production systems in the states of Mato Grosso and São Paulo, by automated collection and analysis systems for CH_4 and CO_2 (Greenfeed).

Several other projects work on subjects that are closely tied to this monitoring and will directly or indirectly collaborate for the continuous improvement of SiM-ABC. Some examples can be cited, for lines of credit financed through ABC, such as projects focused on: analysis of life cycle; survey on carbon stock in biomes/regions and specific national production systems, such as IILPF, for example; and instrumental, low cost and internationally accepted techniques of carbon analysis.

After the beginning of the Multi-institutional Laboratory activities, the SiM-ABC, operational scenario and structuring, it will be possible to make a more complete assessment of the area implemented in the country with the ABC Plan technologies. As it has already been stated, for the area and respective emission reductions to be estimated, the development of digital systems that allow proper collection and management of information is needed, so that there may be a large interaction between the banking system responsible for credit approval under the ABC Program and the research institutions that will monitor its efficiency.

In generating this data, it will be possible to estimate greenhouse gas emissions reduction by technology adopted in the field and, for it to be more accurate, it would be necessary to develop research and obtain emissions factor data specific to each technology and each region of the Brazilian territory.

Methodology for the Emissions Reduction Monitoring

As it has been previously stated, between 2010 and 2012 soil samples were collected throughout the national territory in integrated systems Crop-Livestock-Forest and pastures at different levels of degradation. The Reports of the ABC Observatory FGV (2013) proposes, as a monitoring methodology, the follow up of the evolution of soil carbon concentration in some of the funds approved, as indicated below:

> "For pastures, Crop-Livestock-Forest systems and No-Tillage system, it is essential to know the soil carbon stock, in the beginning of the project and to repeat the carbon analysis every 5 years. The knowledge of the carbon stock is the basis for agriculture of low carbon emission". The report proposes that "the monitoring of greenhouse gas emissions reduction in areas under this technology can be done based on satellite imagery, complemented by the survey and establishment of the soil carbon of the previously establish regions".

For the remaining technologies, different monitoring systems are suggested, as follows:

- "Emissions reduction monitoring due to substitution of nitrogen fertilization for Nitrogen Biological Fixation, will be based on indicators of the sale of inoculants, given mainly by the private sector";
- "Monitoring of the dynamic area with planted forests in Brazil will be done, mainly, with the use of data of the statistic annals of the Brazilian Association of the Producers of Planted Forests (http://www.abraflor.org.br) and information from its associates, and also with annual information of the Sectional Chamber of Forests of the Ministry of Agriculture";
- Animal waste treatment monitoring: this monitoring shall be done, mainly, by means of bank contracts for the installation of biodigestors and its capacity, as well as the reports on the generation of biogas and the quantity of organic compounds generated by the process of composting.

Except for biological nitrogen fixation and the animal waste treatment the Report of the ABC Observatory FGV (2013) also proposes that satellite georeferencing images can be used with minimum resolution of 20 m.

According to the document from the ABC Observatory, "with the indicators cited it will be possible to estimate the soil carbon stock and the vegetal biomass in these areas. Here we use the emissions factors determined by research projects in institutions and Universities". In the monitoring proposal made for SiM-ABC described in this document, the objectives are to adopt these initial methodologies proposed by the ABC Observatory Report FGV (2013) and gradually add new methodologies for a more complete monitoring, structured and strong of the ABC Plan as a whole.

For that to happen, it is necessary to establish a participative environment between the LiM-ABC, with scientific foundations, that allows to provide alternative methodologies to monitor the emissions reduction yearly, which would be operationally implemented towards the configuration of the SiM-ABC definitive scenario proposed, always considering the continuous evolution and improvement of the system, and its many different elements that have already been described. We highlight, therefore, the necessity of continuous improvement of data collection and activities and emissions factors related to the ABC technologies, by region (or state).

Data collection

In order to have efficient data collection and to allow a strong emissions reduction monitoring system there must be an efficient information network. This work will be the responsibility of the Multi-institutional Laboratory of Climatic Changes and Agriculture, which shall establish the guidelines and coordinate the works.

At first, until a specific and more efficient digital system is developed, it is necessary for the Federal Government to intervene in order to have the banking system, responsible for the concession of fund under the ABC Program, provide the minimum information described in this scenario. With this information, it is possible to begin long-term property monitoring. This monitoring will be carried out by high definition satellite imagery. New images must be acquired every two years. We suggest the use of 2010 rapid-eye images acquired by the Ministry of Environment as an essential item for establishing the referential basis and monitoring of the use of technologies in Brazil.

For the creation of a greenhouse gas emissions database, we need a joint effort with several research institutions and universities that work on the subject. We propose that LiM-ABC coordinate, together with MAPA, a large compilation of emissions data, published in indexed magazines and in two editions of the National Greenhouse Gas Inventory (2004 and 2010). Aside from the compilation of data already published, we propose an interaction of this database with the Third Greenhouse Gases National Inventory (to be published in 2014) and with the projects currently underway at Embrapa (*Pecus*—greenhouse gas emissions in livestock production; *Fluxus*—greenhouse gas emissions in the crop production; and *Saltus*—greenhouse gas emissions in native and planted forests). For this, there must be a negotiation between MAPA, the Ministry of Environment, Ministry of Science, Technology and Innovation and Embrapa.

We suggest the creation of this database with the assistance of Embrapa Agricultural Informatics or another outsourcing company, with funds from the Federal Government (MAPA, Ministry of Science, Technology and Innovation, MMA and BNDES).

The Embrapa projects mentioned above work towards generating knowledge of greenhouse gas emissions in the national production systems, and possibly develop specific emissions factors for the country. However, due to the large scale of the project and difficulty of financing, we propose that resources are made available by the Federal Government for the development of emissions factors specific to the ABC technologies, highlighting the integration system Farming-Livestock-Forest, Recovery of Degraded Pastures and No-Tillage.

For the Planted Forest technology and Animal Waste Management, beside the above mentioned actions, we recommend the use of methodologies of the CDM projects already approved during the UNFCCC, as applicable.

Uncertainties

With the development of an efficient digital system for data collection that may support the banking system in this task, and with its alignment with the emissions reduction monitoring system, there is a significant reduction in the monitoring uncertainties. With the acknowledgement of the areas where these technologies were adopted with funds from the ABC program and with the establishment of factors, indicators and operationally optimized methodologies, it is possible to have a long-term high definition monitoring. With this specific emissions factors development to the regions of Brazil (which are to be published in indexed magazines and/or approved by the Intergovernmental Panel on Climate Change Emission Factor Database—EFDB/IPCC and/or used in the National Inventory), the uncertainties of the monitoring will also be significantly reduced (in comparison to the emissions data available today on the text of the ABC Plan).

We encourage the use of CDM projects methodologies already approved at UNFCCC, since they already have monitoring systems, and also the development of new methodologies and CDM projects in Agriculture (especially with the ABC Plan technologies) and their submission to UNFCCC.

Methodological consistency

The emissions reduction monitoring must follow the guidelines accepted internationally (today, from IPCC) using, whenever possible, the most up-to-date documents ('Revised 1996 IPCC Guidelines for National Greenhouse Inventories'—Guidelines 1996 and '2006 IPCC Guidelines for National Greenhouse Gas Inventories'—Guidelines 2006—after its approval), and must detail the principles of Transparency, Accuracy, Consistency, Comparability and Completeness (TACCC). The use of these methodologies is important for the data monitoring to be aligned to the Greenhouse Gas National Inventory.

Guarantee and quality assurance—QA

A quality assurance system must be developed with the aid of specialized technical consultants, in accordance to existing regulations for each type of data collection (analysis of satellite imagery, local greenhouse gases and soil carbon measurements, CDM project methodology, etc.).

Quality assurance (QA) must be based on a system of routine technical activities implemented by the team, who will develop the monitoring system as they go. QA activities must include technical revisions, checking the precision of the system and the use of standardized proceedings approved for the emissions reduction calculation.

It must consist of a system of revision and external audition procedures, carried out by people who are not involved in the development of the monitoring system. Quality assurance must be an independent and objective revision process, developed by third parties in order to evaluate the efficiency of the QA system and the concepts of transparency, accuracy, integration and representativeness of the system.

Report and verification

The data dissemination regarding the emissions reduction monitoring shall be a responsibility of the Ministry of Environment, together with Casa Civil (Staff House of the Presidency of the Republic) and each Ministry responsible for the Sectorial Plan implementation (such as ABC Plan, MAPA and MDA).

We suggest that this broadcasting should be made in the form of reports available on the Ministries' websites. The official announcement will take place, at the international level, following the rules agreed under the UNFCCC and should be done by the Brazil's focal point to the Convention (Ministry of External Relations). This will be an official publication of the country and should include monitoring of all actions performed for the attainment of such national voluntary commitment. However, it is encouraged that each sector disclose the partial results, as information for internal purposes and to aid orientation of domestic public policy.

The verification of data in emissions reduction must be done according to IPCC criteria, by specialists in the field and made accessible to the public, before the official broadcast on the websites of the Ministries involved in each Sectorial Plan.

Conclusions and Recommendations (proposals for improvement)

With the structuring of the SiM-ABC definitive scenario and with its continuous evolution and improvement, we hope to provide to the country the necessary resources to make its efforts to reduce the contribution from the agricultural sector on the global average temperature increase recognized and accepted internationally within the overall goal of keeping warming below 2°C.

Both the ABC Plan and its monitoring system have some vulnerable points that can be improved. With the purpose of reaching these goals by collaborating towards the continuous improvement of the Plan and SiM-ABC, we list current vulnerabilities, as well as suggestions for improvement.

Vulnerability and Suggestions for Improvement of the ABC Plan and its Monitoring

Monitoring:

- Does not predict the development of local emissions studies and the development of emission factor specific to the country;
- Does not account for the alignment of the monitoring actions with the National Inventory and the Annual Estimates;
- Lack of adherence to CDM;
- A gradual strategic adaptation to the international regulations regarding greenhouse gas emissions inventory is necessary, so as to have full integration by 2020.

Technological transfer:

- Lack of support to the Technical Assistance and Rural Extension (TARE) companies in the states;
- Low level of training for the TARE technicians about the Plan technologies;
- Low reach to farmers;
- Low rates of adoption of the Plan's technologies by the farmers.

Financing and credit access:

- Low interaction between the banks and the National Executive Commission/ State Management Groups of ABC;
- Low level of knowledge among the bank employees regarding the works of the ABC Plan/Program;
- Difficulty in broadcasting/adopting funds;
- Extremely low rates of adoption by the small farms to the ABC Program.

Commercial component:

- There is a need to develop marketing strategies for the products and productive chains associated with the technological implementation promoted by the ABC Program;
- Development of prospective studies of the addition and differentiation of products in national and international markets;
- Constant evaluation of the performance of the Brazilian program compared to the commercial partners in South America and other markets.

Alternatives to monitoring

Strength in the process of development of National Inventories and Annual Estimates:

- Need to develop emission factors, and their approval within Emission Factor Database from EFDB/IPCC and its use in national documents;
- Development and use of Tier 2 and 3 methodologies;
- Partnerships between the Government and Educational and Research Institutions.

Research project and interaction of the research with the National Inventory and annual estimates:

- Embrapa projects on greenhouse emissions in livestock (*Pecus*), crops (*Fluxus*) and native and planted forest areas (*Saltus*);
- Embrapa Project/MMA—monitoring of greenhouse gas emissions in iLPF systems in the state of Mato Grosso;
- Embrapa Project/MMA—enteric methane emissions in integrated production systems in the states of Mato Grosso and São Paulo;

- Embrapa Project of continuous improvement of the National Greenhouse Gas Emission Inventory under evaluation at Embrapa. Needs external support;
- Project of CDM methodology development for Agriculture—needs partnership.

Other methodologies:

- More efficient widespread programs of technology transfer and training of technicians and farmers and more complete results verification systems/efficiency of these programs;
- Widespread broadcast of the existing methodologies and development of the CDM methodologies;
- Added strength to the agricultural census (IBGE) with questions about the ABC Plan;
- Comparison of the national mitigation programs with the methodologies of other countries (Exhibit I)—partnerships with research institutions abroad (Labex Embrapa).

Reference

Fundação Getúlio Vargas (FGV). Observatório Agricultura de Baixo Carbono (ABC). Agricultura de Baixa Emissão de Carbono: A evolução de um novo paradigma. 2013. Available at http://www.observatorioabc.com.br/agricultura-de-baixa-emissao-de-carbono-a-evolucao-de-um-novo-paradigma?locale=pt-br

[1] Researcher at Embrapa Agricultural Informatics, Campinas/SP – Brazilian Agricultural Research Corporation (Embrapa).

[a] Email: giampaolo.pellegrino@embrapa.br

[b] Email: ary.fortes@embrapa.br

[c] Email: eduardo.assad@embrapa.br

[2] Embrapa Soils, Rua Jardim Botânico, 1024. Rio de Janeiro/RJ.
Email: renato.rodrigues@embrapa.br

[3] Researcher at Embrapa Cerrados, Planaltina/DF – Brazilian Agricultural Research Corporation (Embrapa).
Email: luiz.cordeiro@embrapa.br

* Corresponding author

5

Agrosilvopastoral Systems in Brazil

An Agricultural Productive Strategy Based on Green Economy Concepts

Júlio César dos Reis,[1,a,*] *Marcelo Carauta M.M. Moraes*[1,b] and *Renato de Aragão Ribeiro Rodrigues*[2]

Introduction

The adverse effects caused by human beings in the environment (such as the emissions of greenhouse gases—GHG, deforestation, air pollution, water and soil contamination and loss of biodiversity) and their economic, social, political and environmental consequences, indicate that the environmental degradation on a global scale is a reality, and, therefore, a paradigm shift related to the organization of productive activities is needed.

The perception that is gaining relevance in discussion rounds related to the present and the future of the world economy is that existing economic models, based on the intensive use of productive factors and devoted exclusively to the generation of profit

[1] Embrapa Agrossilvipastoril, Rodovia dos Pioneiros - MT 222 - Km 2,5, Caixa Postal - 343, CEP - 78550-970 - Sinop/MT.
[a] Email: julio.reis@embrapa.br
[b] Email: marcelo.carauta@embrapa.br
[2] Embrapa Soils, Rua Jardim Botânico, 1024. Rio de Janeiro/RJ.
Email: renato.rodrigues@embrapa.br
* Corresponding author

at any cost, will not easily allow for the targets set by the international community in the Agenda 21 document,[1] as well as the Millennium Development Goals,[2] to be met.

In this sense, it is noteworthy that the term 'Green Economy' has regularly appeared in the recent rounds of discussions held by international institutions.[3] In general, the concept of 'Green Economy' can be understood as the establishment of an economic system that fosters the increase of the well-being and the reduction of social inequalities over time, having as an essential condition the maintenance of the prevailing environmental conditions (UNEP, 2011a). The pursuit of economic and social objectives should not imply the exposure of future generations to considerable risks or to the scarcity of natural resources.

The organization of an economic system based on the provisions of the 'Green Economy' has as a basic feature the establishment of investments in sectors that develop and/or enhance the natural capital,[4] as well as in activities that reduce environmental and ecological risks, next to being labor intensive—which is an important instrument to generate employment and income.

This chapter focuses on the dimension of sustainable agriculture,[5] and more specifically on integrated Crop-Livestock and Forestry (iCLF) systems. This is due to the fact that agriculture is the main economic sector in most Least Developed Countries and in many developing countries (FAO, 2014; World Bank, 2011) and employs about 1.3 billion people worldwide (CEPAL, 2011; FAO, 2011; UNEP, 2011a). Nonetheless, in recent years this production activity has faced the challenge to constantly increase food supply and, at the same time, preserve the available environmental resources. Accordingly, models for the organization of the agricultural production that are based on the cornerstone 'production/productivity increase and environmental preservation' gain momentum.

It is in this context that the proposal to organize the production system based on the integration model C-L-F emerges. This system has as its basic principles the sustainable production through the integration of agriculture, livestock management and forestry activities, carried out in the same area, as intercrops, in succession or rotation, seeking synergistic effects between the components of the agro-ecosystem, allowing for the environmental appropriateness, the appreciation of the human being and the economic viability (Graziano da Silva, 1999; Furtado, 1980; 2000; Balbino et al., 2011; Macedo, 2009; Nair, 1991).

---

[1] Established at the Rio Summit in 1992. For more details on the Agenda 21 targets see: Agenda 21 Document, of the United Nations Conference on the Environment and Development, available at: http://www.un.org/esa/dsd/agenda21/index.shtml. Accessed on July 10th, 2011.

[2] Established at the World Summit on Sustainable Development, held in Johannesburg in 2002. For more details on the targets see: World Summit on Sustainable Development, available at http://www.johannesburgsummit.org/. Accessed on July 7th, 2011.

[3] According to UNEP's (2011) Beyond United Nations Environment Programme, the Green Economy Initiative is one of the nine UN-wide Joint Crisis Initiatives, launched by the UN System's Chief Executives Board in early 2009. In this context, the initiative includes a wide range of research activities and capacity-building events from over 20 UN agencies, including the Bretton Woods Institutions, as well as an Issue Management Group on Green Economy, launched in Washington, DC, in March 2010.

[4] Understood here as composed by biodiversity and the biomes.

[5] For this work, the following elements are considered components of the agricultural sector: cropping, livestock management and the planting of forests.

Therefore, this chapter aims to discuss the proposed use of the integrated model of agricultural and livestock production based on the concept of iCLF, as well as its alignment with the propositions related to the 'sustainable agriculture' aspect included in the proposed restructuring of the production system based on the concept of 'Green Economy'.

Furthermore, the potential gains and benefits that this production technique may yield are identified throughout this chapter, based on general results from initial experiments. The choice of Brazil as the country, was based on its economic and productive relevance in the world's agricultural and livestock market: Brazil is one of the main exporters of agricultural and livestock products, being a market leader in several segments, combining the fact that it is one of the few countries in the world that still have large areas of natural forests (FAO, 2011; MAPA, 2014).

Besides presenting the proposed organization of the agricultural and livestock production system in an integrated manner, this chapter seeks to identify the links of this proposal with the guidelines contained in the document 'Brazilian Agenda 21: Grounds for Discussion' (*Agenda 21 Brasileira: Bases para Discussão*),[6] as well as with the concept of 'Green Economy'. Moreover, it is intended to demonstrate the alignment of the iCLF model with the guidelines and directives contained in the ABC Plan.[7]

In order to fulfill a such goal, four sections follow this introduction: The first (section 2) discusses the main guidelines and implications established by the concept of 'Green Economy'—particularly aspects related to sustainable agriculture—and presents the main points of the ABC Plan; the second (section 3) presents the organization perspective of the agricultural production system based on the precepts of iCLF, taking into account the guidelines previously discussed and using some results found in the literature; the third (section 4) presents some perspectives on the iCLF model in order to provide it as an alternative to the current agricultural production model; and finally the fourth (section 5) includes some concluding remarks.

Green Economy Initiative and the Relevance of the Agricultural Sector

The development path of different countries shows that some of them have achieved high levels of economic and social development. However, in most cases, this result was accompanied by high environmental liabilities: (i) large scale emissions of greenhouse gases, (ii) air pollution, (iii) degradation of natural resources, especially the pollution of water resources, (iv) deforestation, (v) fragmentation of ecosystems, (vi) soil erosion, (vi) changes in soil physical properties, and (vii) the extinction of species.

Other countries also show similar environmental results, however they did not achieve the same level of economic and social development. In other words, the development model widely adopted, based on the provisions of modern

[6] This document was derived from the text elaborated in the Rio Summit's discussion round in 1992.

[7] ABC Plan is the Sectorial Plan for Mitigation and Adaptation to Climate Change for the Consolidation of a Low Carbon Economy in Agriculture, in order to provide the conditions so that the objectives of reducing GHG emissions proposed by Brazil as a voluntary commitment under the United Nations Framework Convention on Climate Change (UNFCCC).

industrialization,[8] in spite of its economic results, intensified the impacts of human actions on the planet in such a way that we now see natural disasters of unprecedented magnitudes in the human history (Graziano da Silva, 1999; Furtado, 1980; 2000).

As a result of this process, there are currently significant changes in the climate and environmental conditions on a global scale, which are important factors for understanding the economic, social and environmental results and perspectives in many countries. Some examples of such changes are: water and energy scarcity, food shortages, high pollution levels, climate change, global warming and abnormal precipitation patterns, increase in the intensity and duration of dry periods and increased incidence and frequency of extreme events such as hurricanes, storms, warmer summers and increased amount of warmer nights, among many others.

Before these issues, measures have been and are still being taken in order to revisit this development pattern, in an attempt to incorporate issues related to environmental preservation, the reversal of already existing environmental liabilities and the appreciation of activities that respect the environment. The key point of this discussion, in the formulation of the development model to be implemented, is the introduction of the sustainability concept. That is to say that the background of this argument is the construction of a new paradigm for the development process based on the balance between technology and environment, in a way as to preserve society's quality of life and well-being, taking into account factors such as environmental preservation and social justice (Bruntland, 1987).

Thus, this new perspective of the development process assumes societies skill and ability to meet their present needs without compromising the ability of future generations to meet their own needs.[9]

In general, the proposed adoption of a sustainable development model is based on the triad: generation of economic, social and environmental benefits. In this sense, according to the document 'Our Common Future' (1987), also known as the Brundtland Report, the main goals of environmental and development policies derived from the concept of sustainable development are: (i) economic growth resume as a necessary condition to eradicate poverty; (ii) constantly innovate the production systems, making them more efficient, democratic and less raw material and energy intensive; (iii) meet the basic human needs such as employment, food, energy, water and sanitation; (iv) preserve the sources of natural resources; (v) appreciate the technological development and risk management; and (vi) include the environmental dimension into the decision-making process.

It is in this context of recognizing the importance of the environmental impacts of production, of increasing concern about the shortage of natural resources and the legacy of a development pattern based on intensive industrialization, that proposals with the precepts of sustainability came to occupy a prominent position. In this political context, we emphasize the proposal launched in 2008 by the United Nations Environment Programme (UNEP) named Green Economy Initiative (GEI) (UNEP, 2009a).

[8] This development model is characterized by enhanced use of technological innovation seeking differential profit, the substitution of labour force by capital, the full use of natural resources, the distribution of environmental costs and the spatial concentration of production and income.

[9] For more details, see: Report of the World Commission on Environment and Development: Our Common Future. Available at http://www.un-documents.net/wced-ocf.htm. Accessed on June 25th, 2011.

As stated by UNEP, the Green Economy Initiative can be defined as a proposal for the re-organization of the economic system in order to increase human welfare and social equity, and, at the same time, reduce environmental risks and the degradation of the natural capital. Considering these guidelines, a 'Green Economy' is an economy that encourages low carbon activities, that makes efficient use of the available resources and that is socially inclusive, to the extent that it adds value to the structural characteristics and productive capacities of the poorest in order to generate income (UNEP, 2011a,b; 2010a).

In view of GEI's underlying assumptions, it is also important to consider that it is possible and economically feasible to boost income and employment through public and private investments in projects and programs that are intended to reduce greenhouse gas emissions and to improve efficiency in the use of resources as well as in the prevention of the loss of natural capital (UNEP, 2010b,c).

According to the GEI's proposal, these strategic investments would aim to foster funding mechanisms for the re-organization of the infrastructure and institutions, with the purpose of adopting sustainable consumption and production practices. Moreover, it should be mentioned that these investments must be in line with the objectives of the fiscal and tax policy, given a pressing need for political reforms and changes in the regulatory mechanisms—so that these fiscal instruments can be used efficiently.

In this sense, the development path advocated by GEI has as its underlying assumptions the preservation, increase and, when necessary, the recovery of the natural capital, considered a key economic factor for the generation of social, economic and environmental benefits. These aspects are even more relevant for the people who rely on natural resources or who directly depend on natural resources to live (UNEP, 2009a,b).

This reconfiguration of the economic system implies increased participation of activities and products originated from sustainable practices in the production mix. The appreciation of these practices tend to allow for the reduction of poverty through direct transfers, job creation, and also allow for the improvement of the access and flow of goods and services originated from the natural capital (UNEP, 2010a). In this context, the key point of this argument is that the transition to a sustainable economic system is based on overcoming the apparent trade-off between economic growth and investment, on the one hand, and gains from the improvement of environmental attributes and social inclusion on the other hand. The main hypothesis of this argument is that the pursuit of the goals related to the improvement of social and environmental conditions may also provide income growth, economic growth and improved welfare.

Another important point is that, according to the guidelines and underlying assumptions of GEI's proposal, investments in activities and sectors that foster structural changes aimed at producing in a sustainable way can mitigate GHG emissions and reduce the volatility of commodity prices (UNEP, 2009a,b). Still, there is the assumption that public sector investments can help the most vulnerable to adapt to environmental changes caused by climate change (UNEP, 2011b).

To this end, these investments should be directed at either making the utilization of scarce natural resources more efficient or helping renew and/or restore them. The cooperative participation of the private sector, particularly through investments aimed at fostering sustainability-oriented activities, is fundamental for the proposed transformation and the establishment of a new paradigm for the development pattern (UNEP, 2010b,c). Strengthening the interdependency between environmental and welfare conditions, between economic and social stability, and still considering the promotion of the profitability of private investments, is the foundation of the GEI proposal for economic development.

It is therefore clear that a change in the trajectory of economic development as preconized by UNEP has economic, social and environmental justifications. In this sense, there are strong justifications and opportunities for intensive participation of the public and private sectors in the promotion of such structural change.

For the public sector, this transition would imply increasing the incentives for the production of items related to sustainable economic activities, the strengthening, expansion and creation of trading conditions of these products, and also institutional restructuring—so that the measures focused on environmental protection are introduced in the contracts signed with the private sector.

With respect to the private sector, this transition would involve a positive response, in terms of contributions and investments, to the institutional, political and structural changes implemented by the public sector (UNEP, 2010b,c). At the same time, in order to achieve these objectives, it is essential that the private sector builds and develops skills and capacities, as well as promotes innovations that allow it to seize the opportunities offered by the proposed implementation of GEI's assumptions (UNEP, 2009a,b).

This holistic approach, as described in GEI's proposal, demonstrates the depth of change in perspective that needs to occur. Moreover, as a planning strategy and as a way to build a transition process to a sustainable model in economic, social and environmental terms, GEI's proposal identifies as key sectors for the beginning of this transformation process those that promote or enable the development and/or the recovery of natural capital, as well as those sectors that are based on activities that reduce the environmental and ecological risks (UNEP, 2011a; 2009a).

Thus, the implementation of GEI's ideas and provisions includes the appreciation and the massive investment in sectors such as: (i) renewable energy; (ii) transport systems with low carbon emissions; (iii) buildings that use energy efficiently; (iv) clean technologies; (v) adequate management of water resources; (vi) improvement of the drinking water supply; (vii) sustainable agriculture; (viii) the responsible management of forest resources; and (ix) improvement of the utilization of fishery resources (UNEP, 2011a). The guidelines in which is pointed the connections between these sectors and the GEI's perspective have as premise the sectors' features in relation to the generation of consistent and positive results in terms of growth of income and wealth, aggregate production, employment generation and poverty reduction.

Within this set of sectors, agriculture has a special role due to its own challenges, in particular the solution of the equation: increase food production while promoting the preservation of natural resources—and also for its direct relation to various social, economic and environmental aspects. In this context, it is worth emphasizing the direct and decisive character of agriculture for the achievement of some of the Millennium Development Goals (MDG)[10] such as eradicating extreme poverty and hunger (MDG 1) and ensuring environmental sustainability (MDG 7).

[10] For more information on the Millennium Development Goals (MDG), see: http://www.un.org/millenniumgoals/.

Next to these structural features of the agricultural sector and its importance as a productive activity on a global scale, according to statistics by the Food and Agriculture Organization (FAO), approximately 2.6 billion people depend on activities related to agricultural production systems (FAOSTAT, 2011). Also according to the FAO, the agricultural sector is the one that most absorbs labor in least developed countries, being in such countries the main employment sector. Another important point is that this sector is the main income generator for poorer individuals.

Taking these aspects into account, World Bank statistics show that the aggregate value of world agricultural production as a percentage of GDP is around 3%, considering the global aggregate production.[11] Nevertheless, such participation level is negatively correlated to the developmental stage of the countries. In the group of developed countries, the average share of agriculture is about 1.5% of GDP, whereas in the least developed countries, this figure is about 30%.

In the Latin American and Caribbean countries, this value is 6%, while in the Middle Eastern countries, agriculture accounts for about 8% of GDP. In the developing countries of East Asia and the Pacific it is 13%, in sub-Saharan Africa it is 16%, and finally, considering the group of emerging economies formed by Brazil, India, China and Russia (BRIC), this corresponds to about 9.5% (World Bank, 2011).

Furthermore, estimates by the World Bank and UNEP indicate that a positive variation in GDP derived from increases in labor productivity in the agricultural sector in developing countries is, on average, three times more likely to increase the income of the poorest quintile in the curve of income distribution than increases in GDP of the same magnitude generated by increases in labor productivity in nonagricultural sectors (World Bank, 2011; UNEP, 2011a).

As currently performed, agriculture may worsen environmental degradation significantly by depleting natural resources and increasing GHG emissions—mainly methane (CH_4) and nitrous oxide (N_2O). This condition turns it into one of the main economic sectors to be considered within the context of the GEI's proposal, at which efforts to consolidate a new productive structure and an innovative development pattern should be directed. These effects are observed both in crop production—in the use of fertilizers and in the management of agricultural areas—as in livestock management, through land degradation and greenhouse gas emissions from cattle raising (Vilela et al., 2008; Mendes and Reis, Jr., 2004).

Furthermore, the use of improper practices and the intensive use of the agricultural production model based on monocultures tend to contribute negatively to the escalation of the degradation process of physical properties—soil density, porosity, structure and consistency—, chemical properties—cation-exchange capacity, acidity, fertility—, and biological properties—mesofauna and soil microorganisms—, as well as to the reduction of crop yields, the increase of the occurrence of weeds, pests, diseases and the increase of the loss of soil biota (Kluthcouski et al., 2003; Martha, Jr. et al., 2006).

[11] According to the methodology used by the World Bank, agricultural production corresponds to International Standard Industrial Classification (ISIC) divisions 1–5 and includes forestry, hunting, and fishing, as well as cultivation of crops and livestock production. Value added is the net output of a sector after adding up all outputs and subtracting intermediate inputs. It is calculated without making deductions for depreciation of fabricated assets or depletion and degradation of natural resources. For more details, see http://data.worldbank.org/about/faq/specific-data-series.

It is also important to consider that improper practices related to soil and the treatment of waste and effluents produced in agriculture can contribute substantially to the pollution of water resources, accelerating erosion, siltation and eutrophication of rivers and causing negative impacts on the quantity, quality and diversity of life in the water.

Taking into account these specificities of the agricultural sector, it is imperative to create and implement actions and public policies specifically aimed at this sector, if adverse effects in terms of contribution to environmental degradation—and generation of environmental liabilities—are to be mitigated. Nonetheless, such policies and actions might enhance the positive effects of agriculture in terms of sustainable food production and income generation, thereby contributing to the reduction of income inequality and poverty (Gasques et al., 2010).

Brazil, through the Presidency's Chief of Staff Office, the Ministry of Agriculture, Livestock and Supply and the Brazilian Agricultural Research Corporation (Embrapa), has shown great interest in promoting practices that mitigate GHG emissions and expand production areas where these technologies are applied. Brazil is also maintaining its historical position of leadership in the fields of Agriculture and Climate Change, through the ABC Plan.

After the 15th Conference of the Parties to the United Nations Framework Convention on Climate Change (COP-15), in order to inform all Parties of the Convention, the Brazilian Government pointed out mitigation actions that the country intended to adopt. The potential greenhouse gas emission reduction resulting from these actions has been estimated to lie in between 36.1 and 38.9% when compared to the country's projected emissions by 2020. To do so, Brazil is implementing several actions, such as reducing deforestation in the Amazon and the Cerrado (Brazilian Savanna), improving energy efficiency and stimulating the widespread adoption of additional sustainable practices in agriculture.

This voluntary commitment was expressed nationally through Law No. 12187 of December 29, 2009; its article 11 provided for the creation of sectorial plans for mitigation and adaptation to climate change in order to consolidate a low carbon economy. These plans should be elaborated in accordance with the National Policy on Climate Change, considering the unique features of each sector, including through the Clean Development Mechanism—CDM and Nationally Appropriate Mitigation Actions—NAMAs. Decree 7390 of December 9, 2010 established that, in order to achieve the national voluntary commitment, several actions will be implemented with the purpose of decreasing between 1.168 and 1.259 billion tons of CO_2e, of a total of 3.236 billion tons of CO_2e projected for the year 2020.

For agriculture, an especial plan was created named 'Sectorial Plan for the Mitigation and Adaptation to Climate Change for the Consolidation of a Low Carbon Economy in Agriculture' (ABC Plan). According to Decree 7390 of December 9, 2010, commitments concerning the agricultural sector refer to actions for the recovery of 15 million hectares of degraded pastures; expansion of the integrated crop-livestock-forest system in four million hectares; expansion of the practice of direct planting (no-till farming) in eight million hectares; expansion of biological nitrogen fixation in 5.5 million hectares of farmland, replacing the use of nitrogen fertilizers; expansion of forest plantation in three million hectares; expanded use of technologies to treat 4.4

million m^3 of animal waste. Therefore, the ABC Plan stands on four main pillars: (i) crop-livestock-forest integration; (ii) direct planting (no-till farming); (iii) biological nitrogen fixation; and (iv) recovery of degraded pasturelands. Planted forests and waste management constitute secondary pillars.

All these measures are equally important at the national and global levels, given the relevance of the Brazilian agricultural sector for the country's economy and for agricultural production around the world. According to FAO (FAOSTAT, 2011) and the Brazilian Ministry of Agriculture (MAPA, 2014), Brazil is the world's leading producer of sugar cane, coffee, orange juice and beans, second largest producer of soy and beef, third largest producer of wheat and chickens, fourth largest producer of corn, grapes, milk and pork, and finally fifth largest producer of cassava and cotton.

Moreover, according to Brazilian Institute for Geography and Statistics (IBGE), in 2009 the agricultural sector accounted for 23% of the country's gross domestic product (GDP).[12] Of this total, agriculture contributed with about 16%, while livestock management with 7% (CEPEA, 2014). In 2010, according to the Brazilian National Food Supply Company (CONAB), approximately 37% of Brazilian exports originated from the agricultural sector. This performance of the agricultural sector is one of the main factors that help explain the observed growth rates of the Brazilian economy in recent years, since this high level of participation in foreign trade has been responsible for maintaining Brazilian trade balance surplus (IBGE, 2014; MAPA, 2014; CONAB, 2014).

The iCLF System in Brazil

The current challenge of the agriculture and livestock production is to combine constant increases in production/productivity with environmental preservation and restoration. The increase in food demand, the lengthy discussions regarding the environmental impacts of agriculture and livestock production and the appreciation of the sustainable aspect of production are structural features that characterize and challenge the contemporary agricultural and livestock activities.

Despite its high productivity, the current production model-based on capital-intensive activities and low diversification—is responsible for very high rates of greenhouse gas emissions, due to significant environmental degradation and its contribution to the intensification of land and income concentration in rural areas (Abramovay, 2000; Balsan, 2006; Campanhola and Graziano da Silva, 2000).

Greater awareness and social mobilization about environmental, social and economic impacts of crop and livestock systems have induced changes in certain management practices and boosted the discussion on the need to adopt production systems that promote the appreciation of environmental aspects, respect for the wellbeing of animals and consider consumers' satisfaction.

The results presented by the production models based on the principles of the 'Green Revolution' were fundamental to meet the growing demand for food in the

[12] According to IBGE methodology, agricultural sector is definite with the sum of the results of economic performance of primary agricultural sector—cropping, livestock management, the planting of forests and fishing—and also the sectors of inputs, industry and distribution.

last 40 years (Balsan, 2006), and to develop numerous technologies that have placed agriculture at the knowledge frontier of scientific research. However, environmental liabilities generated by this system, the intensification of issues related to income and land distribution and growing concerns about environmental impacts of farming activities show that alternatives focused on sustainable food production must be sought and encouraged (Abramovay, 1999; Gasques et al., 2010; Campanhola and Graziano da Silva, 2000).

In order to meet the increased demand for food, producers must expand the total cultivated area, increase productivity or implement a strategy that combines these two alternatives. However, in the current context of sustainability appreciation, the expansion of production through continuous gains in productivity is clearly preferred. The intensification of land use in areas already occupied thereby tends to be one of the most accepted alternatives by the different agents involved with the sustainable development of agriculture and livestock.

However, it is important to note that intensified production systems do not imply excessive or indiscriminate use of fertilizers, but the efficient and rational use of these inputs and compatible technology to maximize profits, using the natural resources in a rational and efficient manner. These issues encourage the search for a new paradigm for sustainable agriculture (Sachs, 1986; Vilela et al., 2008; UNEP, 2009a; 2011a).

It is in this sense that the proposal for the integration of Crop-Livestock-Forestry systems (iCLF) becomes relevant. This production model consists of different production systems of grains, fiber, wood, meat, milk and agro-energy, implemented in the same area, as intercrops, in rotation or in succession, involving the planting of grain, pasture management and associated planted trees (Vilela et al., 2008). The main assumption of iCLF is to be a system of sustainable agricultural production over time (Porfirio da Silva, 2007; Kluthcouski et al., 2003; Martha, Jr. et al., 2007b).

iCLF systems offer the possibility to restore degraded areas by intensifying land use and increasing the complementary and/or synergy effects between the different plant species and cattle raising—therefore raising productivity in a sustainable manner. These systems optimize soil use through grain production in pastureland, and improve pasture productivity due to its constant renewal through the use of residual fertilization of the crop which allows for greater nutrient cycling and enhanced organic matter content in the soil (Trecenti et al., 2008; Vilela et al., 2008; Martha, Jr. and Vilela, 2009).

Furthermore, iCLF systems are regarded as sustainable systems, as they commend: (i) the use of principles of soil and water management and conservation; (ii) the respect for the land-use capacity and agricultural climatic zoning; (iii) the adoption of integrated management of pests, diseases and weeds; (iv) the optimized use of production resources; (v) the employment of no-tillage systems; and (vi) the synergism among crop, livestock and forestry (of which the latter is a central feature) (Kluthcouski et al., 2003; Porfirio da Silva, 2007; Porfirio da Silva et al., 2010).

This new paradigm for the organization of the agricultural production structure is a key instrument for the maintenance of Brazil as a major player in agricultural production in the world and, at the same time, allows for reversing the advance of the environmental degradation process in cultivated areas, especially in the Brazilian *Cerrado*. The inadequate management of economic and environmental resources and inefficient business management explain the fact that currently about 30 million hectares have some degree of degradation (MAPA, 2014).

In degraded pasturelands, the usually observed zootechnical and economic levels are insufficient to ensure the sustainability of livestock management activities (Martha, Jr. et al., 2007b). Moreover, in addition to economic difficulties, pastureland degradation generates environmental problems and may also raise undesirable social impacts such as impoverishment and income concentration in rural areas over time (Martha, Jr. et al., 2007a).

Several recent studies have sought to identify and analyze the economic, social, and environmental benefits provided by integrated crop-livestock production models. In this regard, the work of Alvarenga and Noce (2005), Kluthcouski et al. (2003), Spehar (2006), Porfirio da Silva (2007), Lazzarotto et al. (2009), Martha, Jr. and Vilela (2002), Martha, Jr. et al. (2007b), Martha, Jr. and Vilela (2009), Mendes and Reis, Jr. (2004) and Neto et al. (2010) deserve to be highlighted.

These studies indicate that the iCLF systems enable increase in many aspects related to production and its ecological effects, including: production efficiency, soil conservation and quality, diversification of income for the producer, water conservation, animal performance due to thermal comfort, mitigation of greenhouse gas emissions, adaptation potential to the effects of climate change, and restoration of degraded areas through the intensification of land use. Besides, iCLFs enhance complementary and/or synergistic effects between the different plant species and livestock management, sustainably providing higher production per hectare (Kluthcouski et al., 2003; Trecenti et al., 2008; Lazzarotto et al., 2009).

Furthermore, the use of integrated systems is important because it allows for: (i) increase in biodiversity; (ii) recovery of the forestry component; (iii) generation of shade and reduction of heat or cold; (iv) renovation and/or increase of the organic cycle and that of nutrients, especially when using nitrogen-fixing species; (v) provision of supplemental feeding for animals; (vi) control of the low-growing herbaceous vegetation with a consequent reduction in the use of herbicides and forest fires; (vii) provision of forest-based products with added economic value; (viii) the diversification of forest and livestock products; (ix) improvement of the soil physical and chemical properties; (x) achievement of additional revenue; (xi) erosion control; (xii) increase in water content in the soil; (xiii) provision of better quality pasture during the dry season; (xiv) better use of the labor force in the property and increase of the property value (Schroeder, 1993; Kluthcouski et al., 2006; Porfirio da Silva, 2007).

Next Steps and What Is Expected

In general, the analysis object of studies that seek to evaluate the effects of iCLF systems consists of individual aspects related to the producer and the production; they focus on features and problems found within the property, not highlighting the issues related to its interaction with the environment in which it operates.

Reflecting this analysis approach and according to Martha, Jr. (2010), in general, the focus of the research on integrated systems has been: (i) to assess the potential for reducing the unit production costs resulting from interactions between cultures;

(ii) to evaluate aspects related to risks and vulnerabilities minimization, given the characteristic of production diversification provided by the system; (iii) to assess the variation in profitability among different combinations of integrated production systems; and finally (iv) to assess the potential related to the increase in productivity per unit area. Thus, within a private and individual perspective, the economic benefits of the iCLF system would focus on the possibility of increasing supply at lower costs per production unit.

However, new research fields are gradually becoming part of this discussion, pointing out that the socioeconomic and environmental benefits arising from integrated systems go beyond the individual perspective, and that a broader analysis of these systems—able to evaluate their interactions with the environment—is required.

In this sense, studies that analyze positive socioeconomic and environmental externalities that potentially emerge from the iCLF technology deserve to be highlighted. The possibility of generating employment and income in rural areas, detaining the advancement of the agricultural frontier towards remaining forested areas, mitigating greenhouse gas emissions, increasing carbon stocks in the soil, enhancing the adaptation potential of agricultural systems to future climatic conditions, making use of pesticides and fertilizers more efficiently and reducing water and soil losses are among the main examples (Kluthcouski et al., 2003; 2006; Martha, Jr. et al., 2007b; Martha, Jr. and Vilela, 2009; Martha, Jr. et al., 2010).

However, a more comprehensive analysis targeted at potential effects of integrated systems on their own socioeconomic and environmental externalities, production characteristics, and living conditions of the producer still lacks depth and needs to incorporate more general aspects prior to the implementation and dissemination of this technology. The evaluation of iCLFs economic feasibility illustrates well the issues related to the implementation and to the socioeconomic benefits of integrated systems, especially given producers' socioeconomic characteristics such as income, farm size, level of knowledge and link of this proposal with the recent changes that have occurred in rural areas (especially in relation to changes in the production structure, family structure and the urban-rural relationship). In summary, it is necessary to develop more robust methods which are able to ratify integrated systems' positive effects comparatively to production models based on monoculture, as well as to enable the identification of new effects, such as the possibility of economic exploitation of environmental services.

An important point is that these methods would allow measuring the extent to which the proposed organization of the productive activity, based on the integration of systems, can represent a local development strategy. This last aspect is crucial, because a structural change in the organization of agricultural production, such as the iCLF proposal, requires a large economic, social and institutional apparatus once this production system mobilizes various economic activities simultaneously, requiring a high level of knowledge from the producer.

Moreover, the adoption of an integrated system implies the use of cutting-edge technologies, which opens the possibility to add value to the production and enables greater sectorial integration. The adoption of the iCLF system also induces the promotion of the structural changes necessary to put into practice a local development project based on a process of high and continuous economic growth, oriented to reduce socioeconomic disparities.

Importantly, the perspective of using the iCLF system as a local development strategy allows a close alignment of this policy with the guidelines outlined by the document made by the Commission on Sustainable Development Policies and on the National Agenda 21 (*Comissão de Políticas de Desenvolvimento Sustentável e da Agenda 21 Nacional*) entitled 'Brazilian Agenda 21: Grounds for Discussion' ('*Agenda 21 Brasileira: Bases para Discussão*'), which points out six key elements that threaten the consolidation of sustainability in agriculture. They are: (i) the predominance of the so called 'Green Revolution' standard, intensive in capital and chemical inputs; (ii) the presence of serious environmental liabilities in agribusiness, represented mainly by the high level of soil erosion, degradation of water resources and loss of biological diversity; (iii) foreign scientific and technological dependence, stressed by the 'Green Revolution' and aggravated by the lack of true national innovation; (iv) the dominance in the agricultural sector of a primary-good-export model, based on external demands, generator of environmental and social costs which importing countries are reluctant to incorporate into the final prices; (v) profitability based on the intensive and predatory exploitation of natural and environmental resources; and (vi) highly concentrated land ownership, which tends to become even more intense in many parts.

Still, it is important to note that the proposed use of the integrated system as an instrument of social transformation represents the implementation of the guidelines established by the Brazilian Federal Government aimed at achieving a new development model for the *Cerrado* and Amazon regions. These guidelines are based on social inclusion, reduction of socioeconomic inequalities and respect for cultural diversity, turned to dynamic and competitive economic activities that sustainably generate employment. Among these plans, a few should be highlighted: the Regional Sustainable Development Plan for the impact area of BR-163—the road Cuiabá-Santarém—, the Regional Sustainable Development Plan for Xingu and the Sustainable Amazon Plan (PAS), all coordinated by the Ministry of National Integration.

Final Remarks

The adoption of the Green Economy Initiative (GEI) emerges as a consistent strategy for the implementation of policies that aim to promote sustainable development. The GEI recognizes that the end goal of sustainable development is the promotion of social welfare with respect for the constraints imposed by the environment, including measures to mitigate the negative causes and effects of climate change, energy scarcity and environmental degradation.

However, it is important to note that GEI should not be interpreted as a strategy focused exclusively on the elimination of environmental problems. To the contrary, it should be interpreted as an alternative that aims at promoting sustainable development, boosting welfare and alleviating poverty. And this is a central topic given the current development of many conflicts whose underlying causes relate to issues such as unemployment, income inequality and access to natural resources—such as water and land.

Considering the GEI proposal, as well as the sectors it points out as the most relevant for the implementation of the required process of structural change, the agricultural sector represents an important component of this transformation. The solution of the complex equation of increasing food supply under the constraints imposed by environmental factors poses itself as a major challenge for today's society.

In this context, the iCLF model is one of these alternatives; although restructuring the agricultural production still requires further studies, as mentioned above, its early positive results in economic, social and environmental aspects makes it a concrete possibility to overcome the paradigm of increasing food production through the predatory use of natural resources. Furthermore, the iCLF proposal has as underlying assumptions the generation of income and the promotion of conditions that permit the current population to stay in the rural areas. All these aspects are fundamental for the development of strategies to overcome the poverty conditions of people living in the rural areas.

In this sense, the strategy of organizing the agricultural production based on the iCLF model is perfectly aligned with GEI's assumptions regarding the promotion of sustainable agriculture models and the provision of incentives to their consolidation. Furthermore, considering initial results, this system can strengthen Brazil's leadership position in various segments related to agricultural production, as well as contribute to the establishment of a new paradigm for the organization of the agricultural production system as a whole.

Acknowledgments

The authors would like to thank Embrapa Agrosilvopastoral for providing all the material that was necessary to accomplish this work. Furthermore, we wish to thank colleagues Austerclínio Lopes de Farias Neto, João Flávio dos Reis Veloso, Lineu Alberto Domit, Eduardo da Silva Matos and Luciano Bastos Lopes for their comments and suggestions. In this sense, all possible mistakes are author's responsibility.

References

Abramovay, R. 1999. Funções e medidas da ruralidade no desenvolvimento contemporâneo.

Alvarenga, R.C. and Noce, M.A. 2005. Integração lavoura-pecuária. Embrapa Milho e Sorgo.

Balbino, L.C., Barcellos, A. de O. and Stone, L.F. 2011. Marco referencial: integração lavoura-pecuária-floresta, 1ª ed. Luiz Carlos Balbino, Alexandre de Oliveira Barcellos, Luís Fernando Stone, Brasília, DF—Embrapa Sede.

Balsan, R. 2006. Impactos decorrentes da modernização da agricultura brasileira. CAMPO-Territ. Rev. Geogr. Agrár. 1.

Bruntland, G.H. 1987. Our Commom Future. The World Commission on Environment and Development. Oxford: Oxford University Press. New York. NY.

Campanhola, C. and Graziano da Silva, J. 2000. Diretrizes de políticas públicas para o novo rural brasileiro: incorporando a noção de desemvolvimento local. pp. 61–89. *In*: Campanhola, C and Graziano da Silva, J. (ed.). O Novo Rural Brasileiro: Políticas Públicas. Volume 4. Capítulo 3. Embrapa Meio Ambiente. Jaguriúna - São Paulo.

CEPAL. 2011. Comisión Económica para América Latina y el Caribe. CEPAL [WWW Document]. URL http://www.cepal.org/(acccessed 1.7.11).

CEPEA. 2014. Centro de Estudos Avançados em Economia Aplicada—CEPEA/ESALQ/USP [WWW Document]. URL http://www.cepea.esalq.usp.br/(accessed 4.2.14).

CONAB. 2014. Companhia Nacional de Abastecimento [WWW Document]. URL http://www.conab.gov.br/(accessed 4.3.14).

FAO. 2014. Food and Agriculture Organization of the United Nations [WWW Document]. URL http://www.fao.org/home/en/(accessed 4.2.14).

FAO. 2011. Food And Agriculture Organization of the United Nations [WWW Document]. URL http://www.fao.org/home/en/(accessed 5.7.11).

FAOSTAT. 2011. FAOSTAT [WWW Document]. URL http://faostat3.fao.org/faostat-gateway/go/to/home/E (accessed 4.2.14).

Furtado, C. 1980. Pequena introdução ao desenvolvimento—um enfoque interdisciplinar. Companhia Editora Nacional, São Paulo, SP.

Furtado, C. 2000. Introdução ao desenvolvimento: enfoque histórico-estrutural. Paz e Terra, São Paulo, SP.

Gasques, J.G., Vieira Filho, J.E.R. and Navarro, Z. (eds.). 2010. A Agricultura Brasileira: desempenho, desafios e perspectivas. IPEA, Brasília, DF-IPEA.

Graziano da Silva, J.F. 1999. O novo rural brasileiro. Universidade Estadual de Campinas, Instituto de Economia.

IBGE. 2014. Instituto Brasileiro de Geografia e Estatística [WWW Document]. URL http://www.ibge.gov.br/home/ (accessed 4.2.14).

Kluthcouski, J., Aidar, H., Cobucci, T., Stone, L.F., Thung, M.D.T., Balbino, L.C., Silva, C.C. da and Oliveira, F.R. de. 2006. Integração lavoura-pecuária: estudo de caso vivenciado pela Embrapa Arroz e Feijão. pp. 277–330. *In*: Paterniani, Emesto (ed.). Ciência, Agricultura E Sociedade. Embrapa Informação Tecnológica, Brasília, DF.

Kluthcouski, J., Stone, L.F. and Aidar, H. 2003. Integração lavoura-pecuária. Embrapa Arroz e Feijão, Santo Antônio de Goiás.

Lazzarotto, J.J., dos Santos, M.L., de Lima, J.E. and de Moraes, A. 2009. Volatilidade dos Retornos Econômicos Associados à Integração Lavoura-Pecuária no Estado do Paraná. Rev. Econ. E Agronegócio 7.

Macedo, M.C.M. 2009. Integração lavoura e pecuária: o estado da arte e inovações tecnológicas. Rev. Bras. Zootec. 38: 133–146.

MAPA. 2014. Ministério da Agricultura, Pecuária e Abastecimento [WWW Document]. URL http://www.agricultura.gov.br/(accessed 4.2.14).

Martha, Jr., G.B., Vilela, L. and Santos, D. de C. 2010. Dimensão Econômica da Soja na Integração Lavoura-Pecuária. Ata XXXI Reunião Pesqui. Soja Reg. Cent. Bras., Série Documentos 324—Embrapa Soja 3, 80–88.

Martha, Jr., G.B., Barcellos, A. de O., Vilela, L. and Sousa, D.M.G. 2006. Benefícios Bioeconômicos e Ambientais da Integração Lavoura-Pecuária (Série Documentos No. 154). Embrapa Cerrados, Planaltina, DF.

Martha, Jr., G.B. and Vilela, L. 2002. Pastagens no cerrado: baixa produtividade pelo uso limitado de fertilizantes. Sér. Doc. 50—Embrapa Cerrados 32 p.

Martha, Jr., G.B. and Vilela, L. 2009. Efeito poupa-terra de sistemas de integração lavoura-pecuária. Sér. Comun. Téc. 164—Embrapa Cerrados 4 p.

Martha, Jr., G.B., Vilela, L., Barcellos, A. de O., Sousa, D.M.G. and Barioni, L.G. 2007a. Pecuária de corte no Cerrado: aspectos históricos e conjunturais. pp. 17–42. *In*: Martha Júnior, G.B., Vilela, L. and Sousa, D.M.G. de (ed.). Cerrado: Uso Eficiente Decorretivos E Fertilizantes Em Pastagens. Embrapa Cerrados, Planaltina, DF.

Martha, Jr., G.B., Vilela, L. and Maciel, G.A. 2007b. A prática da integração lavoura-pecuária como ferramenta de sustentabilidade econômica na exploração pecuária. Presented at the Simpósio de Forragicultura e Pastagens, Lavras, MG—Universidade Federal de Lavras (UFLA) pp. 367–391.

Mendes, I. de C. and Reis, Jr., F.B. 2004. Uso de parâmetros microbiológicos como indicadores para avaliar a qualidade do solo e a sustentabilidade dos agroecossistema. Sér. Doc. 112-Embrapa Cerrados 34 p.

Nair, P.K.R. 1991. State-of-the-art of agroforestry systems. For. Ecol. Manag. 45: 5–29.

Neto, S.N. de O., Vale, A.B., Nacif, A. de P., Vilar, M.B. and Assis, J.B. 2010. Sistema Agrossilvipastoril: integração lavoura, pecuária e floresta, 1ª ed. Sílvio Nolasco de Oliveira Neto ... [et al.] organizador, Viçosa, MG.

Porfirio da Silva, V. 2007. A integração "lavoura-pecuária-floresta" como proposta de mudança do uso da terra, in: Novos Desafios Para O Leite No Brasil. Juiz de Fora, MG—Embrapa Gado de Leite pp. 197–210.

Porfirio da Silva, V., Medrado, M.J.S., Nicodemo, M.L. and Dereti, R.M. 2010. Arborização de pastagens com espécies florestais madeireiras: implantação e manejo. Embrapa Florestas, Colombo, PR.

Sachs, I. 1986. Ecodesenvolvimento: crescer sem destruir. Edições Vértice, São Paulo, SP.

Schroeder, P. 1993. Agroforestry systems: integrated land use to store and conserve carbon. Clim. Res. 3: 53–60.

Spehar, C.A. 2006. Conquista do Cerrado e consolidação da agropecuária. pp. 195–226. *In*: Paterniani, Emesto (ed.). Ciência, Agricultura E Sociedade. Embrapa Informação Tecnológica, Brasília, DF.

Trecenti, R., Oliveira, M.C. and Hass, G. 2008. Integração lavoura-pecuária-silvicultura: boletim técnico. Sér. Bol. Téc.-Minist. Agric. Pecuária E Abast. Secr. Desenvolv. Agropecuário E Coop. 54 p.

UNEP, U.N.E.P. 2009a. Rethinking the Economic Recovery: A Global Green New Deal.

UNEP, U.N.E.P. 2009b. Global Green New Deal Policy.

UNEP, U.N.E.P. 2010a. Developing Countries Success Stories.

UNEP, U.N.E.P. 2010b. Driving a Green Economy Through Public Finance and Fiscal Policy Reform.

UNEP, U.N.E.P. 2010c. A Brief For Policymakers on the Green Economy and Millennium Development Goals.

UNEP, U.N.E.P. 2011a. Towards a Green Economy: Pathways to Sustainable Development and Poverty Eradication.

UNEP, U.N.E.P. 2011b. Why a Green Economy Matters for the Least Developed Countries.

Vilela, L., Martha, Jr., G.B., Marchão, R.L., Guimarães, Jr., R., Barioni, L.G. and Barcellos, A. de O. 2008. Integração Lavoura—Pecuária. pp. 932–962. *In*: Savanas: Desafios E Estratégias Para O Equilíbrio Entre Sociedade, Agronegócio E Recursos Naturais. Fábio Gelape Faleiro, Austeclinio Lopes de Farias Neto, Plana ltina, DF.

World Bank. 2011. The World Bank [WWW Document]. URL http://data.worldbank.org/ (accessed 3.7.11).

6

Soil Conservation and Carbon Sequestration

Agroforestry Systems as a Promising Alternative

Eduardo da Silva Matos,[1,*] *Eduardo de Sá Mendonça,*[2] *Dirk Freese*[3] and *Marcela Cardoso Guilles da Conceição*[4]

Introduction

One of the challenges that modern agriculture has to face is improving cropping systems and soil management strategies in order to remove significant amounts of CO_2 from the atmosphere (Chabbi and Rumpel, 2009) and promote the conservation of water resources. The amount of C loss and input, in a specific agrosystem, will determine its direction to sustainability or degradation.

The use of appropriate agricultural practices is essential for maintenance of water resources, since these practices reduce the loss of soil due to erosion and leaching and maintains favorable content of organic matter in the soil.

[1] Embrapa Agrosilvopastoral, PO Box 343, Sinop, MT 78550-970, Brazil.
Email: eduardo.matos@embrapa.br

[2] Department of Plant Production, Federal University of Espirito Santo, Alto Universitario s/n, Alegre, ES 29500-000, Brazil.
Email: eduardo.mendonca@ufes.br

[3] Soil Protection and Recultivation, Brandenburg University of Technology, Konrad-Wachsmann-Allee 6, D-03046 Cottbus, Germany.
Email: freese@tu-cottbus.de

[4] Department of Environmental Geochemistry, Fluminense Federal University, Rua Outeiro São João Batista s/n, Niterói, RJ 24020-007, Brazil.
Email: marcelaguilles.clima@gmail.com

* Corresponding author

Different types of management in agriculture affect the amount of soil organic matter, soil structure, soil depth, and water and nutrient-holding capacity. One goal of soil quality research is to learn how to manage soil in a way that improves soil function, since soils respond differently to management depending on the inherent properties of the soil and the surrounding landscape (United States, 2013).

Many soil properties impact soil health, but organic matter deserves special attention. It affects several critical soil functions, can be manipulated by land management practices, and is important in most agricultural settings across the world. Because organic matter enhances water and nutrient-holding capacity and improves soil structure, managing for soil carbon can enhance productivity and environmental quality, and can reduce the severity and costs of natural phenomena, such as drought, flood, and disease (Fig. 6.1). In addition, increasing soil organic matter levels can reduce atmospheric CO_2 levels that contribute to climate change (United States, 2013).

Especially important within these contexts are the agricultural practices that reduce the CO_2 emission to the atmosphere by incorporating part of it in the vegetation, as aerial and root biomass, and in the soil, as organic matter. The means by which agricultural and forestry practices can reduce greenhouse gases emissions are: (1) avoiding emissions by maintaining existing C storage in trees and soils; (2) increasing C storage by tree planting, conversion from conventional to conservation tillage practices on agricultural lands; and (3) substituting bio-based fuels and products for fossil fuels, such as coal and oil, and energy-intensive products that generate greater quantities of CO_2 when used (United States, 2014).

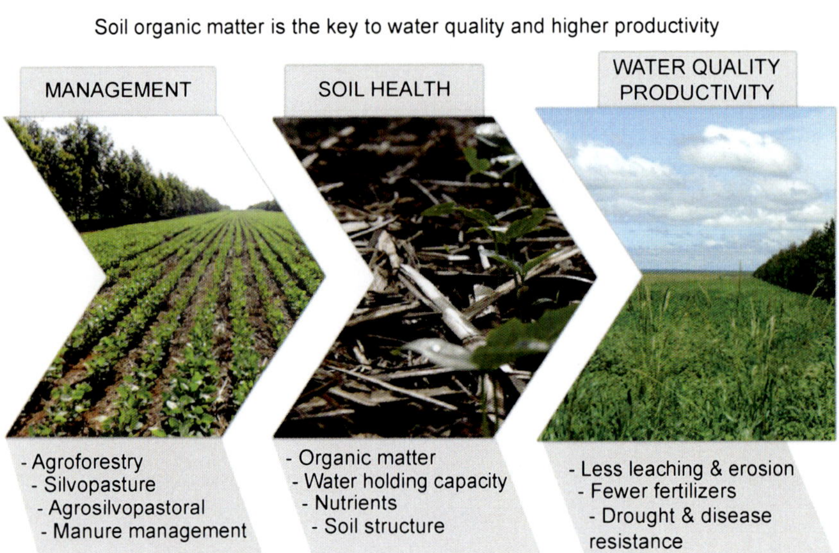

Figure 6.1. Adapted of United States (2013) (Illustration: Keyle B. Menezes).

Low Carbon Agriculture

In tropical and subtropical regions, the degradation of the organic fraction of the soil in inappropriate management conditions is rapid and is accompanied by a degradation process of the chemical, physical and biological conditions of the soil. In these regions, appropriate management for soil conservation and crop yield should be premised on the use of methods of preparation with minimal or no soil tillage and rotation/succession of crops including legumes and crops with high production systems plant residues.

Conservation tillage practices or conservation agriculture or low C agriculture systems are characterized by minimal soil disturbance, constant soil coverage and crop rotation. Especially important in this context are the management systems, which combine production of tree or wood products, agronomic crops and/or pasture in the same area. Agroforestry systems have provided a sustainable alternative to shifting agricultural and single crop systems because of its potential to restore degraded soils contributing to increase short-term C storage in vegetation, soil and biomass products (Oelbermann et al., 2006b; Nair et al., 2009).

Agroforestry systems have been used in temperate and tropical zones in order to improve economic, ecological, and environmental benefits of farmlands (Tapia-Coral et al., 2005; Oelbermann et al., 2006b; Reynolds et al., 2007). As in tropical regions, the objectives for establishing agroforestry systems in temperate regions are similar which correspond to the production of tree or wood products, agronomic crops or pasture, livestock, and improvement of crop quality and quantity (Jose et al., 2004). Management practices to attain these objectives have included enhancement of microclimatic conditions; improved utilization and recycling of soil nutrients; improved soil and water quality; provision of favorable habitats for plant, insect or animal species; soil stabilization (reducing erosion in hilly regions); and protection from wind and storm, among others (Fig. 6.2).

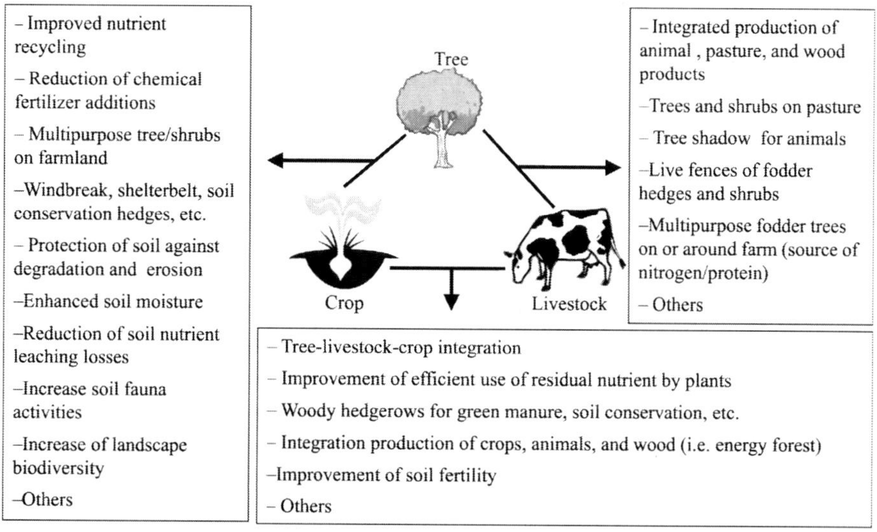

Figure 6.2. The interaction of tree, crop, and livestock under agroforestry systems.

Agroforestry systems have shown high potential to improve soil physical, chemical and biological properties through accretion and decomposition of organic material derived from litterfall and roots (Haile et al., 2008). Soil fertility improvement is a distinct possibility in agroforestry systems, especially important in the tropics where the high weathered soils present, in general low natural fertility associated with deep profiles, clay mineralogy retrained by oxyhydroxides, high aluminum saturation and P adsorption capacity (Cardoso et al., 2003). In these areas, the management of different plant species with distinct capacity to remove or extract nutrients from and to different soil depths could contribute to improve soil fertility and soil quality allowing the sustainable management of nutrient in poor soils.

Silvopasture as an agroforestry practice integrates trees with pasture and livestock production (Fig. 6.3). The trees are managed for high quality timber component or tree products and at the same time provide shade, forage and shelter for livestock. It is well known that trees can improve soil conditions in agroforestry pastureland systems through various processes and mechanisms including reduced soil erosion, increase in soil organic matter content, increased soil N through N_2 fixation, and changes in soil chemical, physical and biochemical properties (Dagang and Nair, 2003). Compared with treeless pastures, silvopastoral systems have a greater potential for C sequestration (Haile et al., 2008) and also to remove or extract nutrients from deeper layers, which could contribute to improve nutrient availability for pasture.

Trees in pastures can also contribute to biodiversity conservation. Nair et al. (2007) suggested that silvopastoral systems provide greater environmental benefits as compared to treeless pasture under similar ecological conditions by increasing C sequestration and nutrient removal. Michel et al. (2007) hypothesized that the incorporation of trees in pastures could enhance P storage in the soil-plant system reducing P losses and consequently water pollution in areas under pasture in Florida. They concluded that the silvopastoral system has a higher capacity to store additional P than treeless pastures. Thus, silvopasture would provide a greater environmental service in regard to water quantity and quality protection, as well as soil quality compared to treeless pastures.

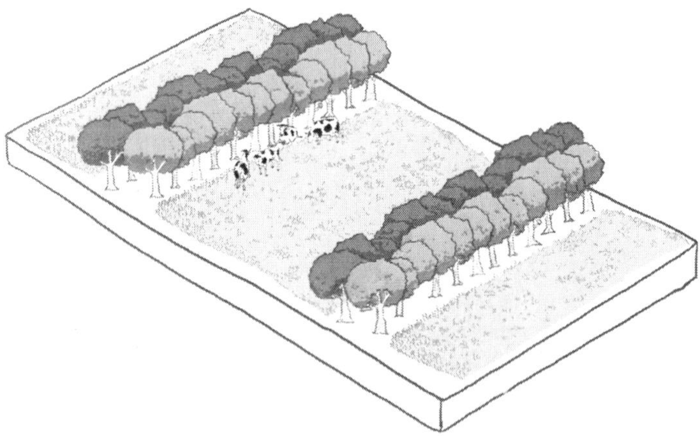

Figure 6.3. Model of a silvopastoral system (Illustration: Keyle B. Menezes).

Alley cropping system is an agroforestry practice in which arable crops are cultivated between tree hedgerows (Fig. 6.4). Alley cropping has been used in temperate zones for the production of trees and crops or livestock (Reynolds et al., 2007), and in the tropics associated with coffee, cocoa and other trees (Malézieux et al., 2009). The presence of woody species in the alley cropping production system has shown to contribute to nutrient cycling, reduction in soil nutrient leaching losses, and stimulation of higher soil faunal activities, soil erosion control, soil fertility improvement and sustained levels of crop production (Kang, 1997).

Alley cropping systems have also come into focus for reclamation of post-mining areas (Gruenewald et al., 2007; Nii-Annang et al., 2009), where the stocks of Soil Organic Matter (SOM) are generally low and soil properties are not suitable enough for plant growth. The increase of SOM in reclamation areas depends on the amount of biomass production and return to the soil, and mechanisms of C protection (Gruenewald et al., 2007). Therefore, the cultivation of fast-growing tree species have been largely used for the production of energy forests showing great benefits for reclamation of post mining areas (Bungart and Hüttl, 2004; Gruenewald et al., 2007). Production of *Robinia pseudoacacia* L. under alley cropping system has received considerable interest in these areas as an alternative to agricultural crops and an additional wood source, while acting as a potential C sink to counterbalance greenhouse gases emissions. According to Gruenewald et al. (2007), in an alley cropping system production, the above-ground biomass accumulation of *R. pseudoacacia* reached 29.8 t ha^{-1} after six years of establishment on reclaimed mine sites in northeastern Germany. The same authors observed values of above-ground biomass accumulation six times less (around 4.5 t ha^{-1}) for *Salix viminalis* L. and attributed these differences to the adaptability of *R. pseudoacacia* to nutrient-poor sandy substrates and low precipitation regime. Due to its high potential of litterfall production and N_2 fixation, *R. pseudoacacia* might contribute to improve soil physical, chemical and biological properties though increasing SOM, converting spoils into productive and sustainable soils.

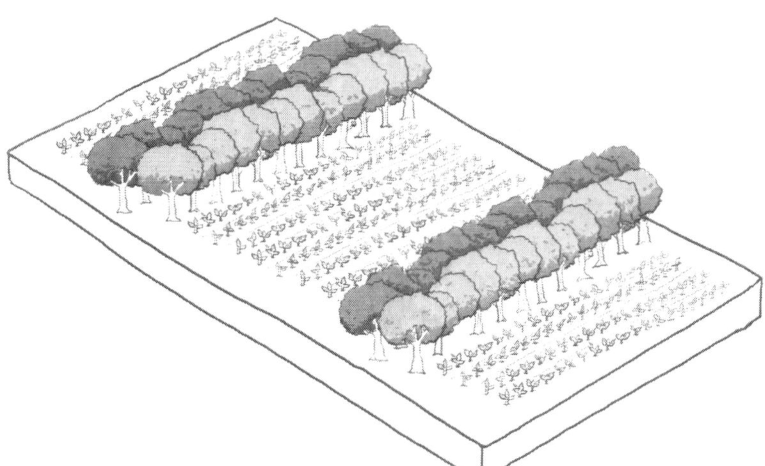

Figure 6.4. Model of an alley cropping system (Illustration: Keyle B. Menezes).

In the humid tropics, alley cropping systems have also been considered as an alternative for adequate soil use (Kang, 1997; Oelbermann et al., 2005; Moura et al., 2010). However, evaluating the effect of alley cropping on dynamics of SOM in Amazonia, Aguiar et al. (2013) showed that 10 years of old alley cropping did not contribute to increase light fraction of organic matter. Furthermore, labile fractions of organic matter decreased with continuous use of alley cropping. This finding shows that more investigations are necessary to better understanding the effects of agroforestry systems in tropical soils. High temperatures and humidity contribute to high decomposition rates. Thus, research should also focus on the quality of inputs provided by tree or annual litterfall and non-tree components of an alley cropping system.

The management system that combines not only two components, but three of them (cropping, livestock and forestry) in the same area, associated with others conservation tillage practices has also been considered as an important alternative to restore degraded soils and increase short-term C storage in vegetation, soil and biomass products (Maciel et al., 2011). Agrosilvopastoral system or integrated crop-livestock-forest system consists in growing forest species simultaneously with commercial crops, although, after crop harvest the area is cultivated with forages for livestock. It is also possible to cultivate forages associated with maize or sorghum at the same time. Hence, after grain harvesting, the pasture is already established between the tree rows, enabling animal grazing (Pacheco et al., 2013). As well as for alley cropping or silvopastoral system, integrated crop-livestock-forest system could also contribute to increasing soil biodiversity, protecting soil against erosion, reducing soil nutrient leaching losses and improving nutrient recycling (Nair, 1985; Maciel et al., 2011; Pacheco et al., 2013). Furthemore, these management systems permit increasing total production, reducing risks and diversifying the cultivations, thus being more economically attractive than monoculture (Dube et al., 2002).

Soil Organic Matter in Different Land Use Systems

Soils contain the largest near-surface reservoir of terrestrial C (Neff et al., 2002). According to Schimel (1995), globally, soil organic matter (SOM) including detritus contains $\approx$ 1580 Gt of C as compared to $\approx$ 610 Gt C in terrestrial vegetation, $\approx$ 750 Gt C in the atmosphere, and 1020 Gt C in the surface ocean. The amount of organic C contained in a particular soil is a function of the balance between the rate of deposition of plant residues on soil and the rate of mineralization of the residue C by soil biota (Baldock and Nelson, 1999). This balance is especially important in high weathered soils, as tropical soils, where SOM plays important role on the sustainability of the agricultural system due to its benefits on soil attributes and plant growth (Maciel et al., 2011). According to Post and Kwon (2000), there are many factors and processes that determine the direction and rate of SOM dynamic. The factors and processes that are important for increasing soil organic C storage include (1) increasing the input rates of organic matter, (2) changing the decomposability of organic matter inputs that increase light fraction in particular, (3) placing organic matter deeper in the soil either directly by increasing belowground inputs or indirectly by enhancing surface mixing by soil organisms, and (4) enhancing physical protection through either intra-aggregate or organo-mineral complexes.

Decreases in SOM leads to a decline in agricultural and biomass productivity, poor environmental quality, soil degradation, and nutrient depletion (Leblanc et al., 2006). Therefore, the combination of agricultural production and environmental services particularly C sequestration, through improved and well managed soil practices, appears to be a good alternative (Palm et al., 2013).

The amount of C sequestered largely depends on the type agroforestry system, and is determinate by environmental and socio-economic factors. Other factors influencing carbon storage in agroforestry systems include tree species and system management (Albrecht and Kandji, 2003). Agroforestry systems play an important role as a C sequestration strategy because of C storage in its multiple plant species and soil as well as its applicability in agricultural lands and in reforestation (Montagnini and Nair, 2004). In soil, this process is accomplished by the increase in labile and recalcitrant forms of C being necessary to investigate the changes in SOM pools and turnover rates in natural and agricultural systems to estimate the fluxes of C between the soils and the atmosphere. According to Dixon (1995), establishment of agroforestry systems would allow the storage of approximately 95 Mg C ha^{-1} in boreal, temperate and tropical regions. In a review, Montagnini and Nair (2004) reported that average C storage by agroforestry is estimated as 63 Mg C ha^{-1} in temperate regions. In the humid tropics, Mutuo et al. (2005) estimated as 70 Mg C ha^{-1}, the potential of agroforestry systems to storage C in vegetation, and up to 25 Mg ha^{-1} in the top 20 cm of soil.

However, the impact of an agroforestry system on soil C sequestration depends largely on the amount and quality of inputs provided by tree or annual litterfall and non-tree components of the system and on the soil properties, such as soil structure and texture (Nair et al., 2009). In the other hand, these residues influence soil aggregation, and enhance levels of SOM providing a pathway for the long-term sequestration of C in soil (Oelbermann et al., 2006a).

Tree-based land-use systems are expected to have higher soil C sequestration potential than most row crop agricultural systems since the tree component in agroforestry systems can be significant sinks of atmospheric C due to their high and long-term biomass stock and extensive root system (Montagnini and Nair, 2004). The establishment of tree-based land use systems modifies above- and belowground productivity, rooting depth and distribution, and the quantity and quality of organic matter inputs compared to sole cropping (Haile et al., 2008).

Evaluating different management systems Haile et al. (2008), hypothesized that compared with open pasture systems, silvopastoral agroforestry systems are likely to enhance soil C sequestration in deeper soil layers. The authors observed that in whole soil, total organic C was increased by 33% in silvopastures near trees and by 28% between tree rows compared to an adjacent open pasture. The increase in whole soil was particularly more significant at lower depths. Nair et al. (2007) also reported higher organic C stocks in deeper profile in a silvopastoral system. Using ^{13}C technique, they observed that the tree C_3 contribution in the whole soil was generally higher throughout the soil profile of the silvopasture as compared to the treeless pasture. In a tropical silvopastoral system, Maia et al. (2007) reported total organic C stocks were around 68.7 Mg ha^{-1} at 0–40 cm depth. In the meantime, an adjacent area under intensive cropping system presented values of C stocks of 44.3 Mg ha^{-1}. The authors attributed the high C stocks under silvopasture to the plant residues derived from pruning, manure and weeding.

After 13 years of alley cropping introduction in southern Canada, total soil C was increased in 8 g kg^{-1} showing that temperate alley cropping systems have the potential to sequester C in soil (Oelbermann and Voroney, 2007). Evaluating the effects of alley cropping systems on the reclamation of a lignite mining area in Germany, Nii-Annang et al. (2009) observed significant increases on soil organic C and N contents after nine years of recultivation with the highest accumulation under hedgerows. The C stocks in the first 0–3 cm layer corresponded to 5.3 and 7.8 Mg ha^{-1} under *Robinia pseudocacia* and *Populus* spp. hedgerows, respectively, and to 1.3 Mg ha^{-1} in the alleys under rye. Furthermore, alley cropping contributed to improve soil microbial properties in response to increased organic matter inputs. Investigating alley cropping system in temperate and tropical regions, Oelbermann et al. (2006a) observed higher soil C stocks under tropical alley cropping (95.3 Mg ha^{-1}) compared to sole crop (61.4 Mg ha^{-1}) at 20 cm depth. They attributed this increase to the higher input of organic material from crop residues and from tree pruning in the alley cropping system.

In general, increase in soil C as an effect of conservation tillage practices is observed in long-term experiments. However, different fractions of SOM have been investigated to understand SOM dynamics under different land-use systems, climate conditions, vegetation and topography. The different fractions of SOM are very heterogeneous in structure and differ in decomposability, recalcitrance and turnover rate. The identification of more sensitive SOM fractions contributes to assess changes caused by management practices and trajectories in soil C pool at early stages (Leifeld and Kögel-Knabner, 2005). SOM fractions, such as microbial biomass carbon and light fraction organic carbon, generally respond more rapidly to changes in land uses than total organic matter contents (Leite et al., 2003; Wu et al., 2003; Leifeld and Kögel-Knabner, 2005) and heavy fraction, which is associated with clay and silt. Light fraction represents a free and noncomplexed carbon pool, which can be physically fractionated by density separation (Roscoe et al., 2001; Sohi et al., 2001; Wu et al., 2003).

Bayer et al. (2010) found that in Cerrado soils, carbon build-up in tillage system occurs preferentially in particulate organic matter protected in soil aggregates, more sensitive than the total SOC to changes in soil management. According to Freitas et al. (2000) for the Cerrado Oxisols, the incorporation of waste in the conventional tillage system favors the storage of C in the labile compartment composed of 'plant debris' in comparison with the conventional system.

Light fraction has been considered as a dynamic fraction that reflects in short-term shifts in SOM storage and turnover rate induced by different management practices (Leite et al., 2003; Maia et al., 2007; Murage et al., 2007). It is affected by land use, vegetation type, climate, soil type, faunal activity, among others. In soil with permanent vegetation light fraction can account for 15–40% of SOM, whereas in long-term cultivated arable soils it usually presents values less than 10% of the SOM in the tilled layer (Christensen, 2001). In northeastern Brazil, Maia et al. (2007) reported values of light fraction C around 7.0 Mg ha^{-1} in soil under a silvopastoral system in the 0–12 cm layer. While not statistically significant, the same authors observed light fraction values around 40% lower under conventional tillage systems and attributed the low contents to soil disturbances during the five years of cropping systems. Investigating the effects of an agroforestry system in central-south Mediterranean Chile, Munõz et al. (2007) observed that light fraction was influenced by the presence of tree component (*Acacia caven*) with values being greater under tree canopy than outside its canopy. The same authors concluded that light fraction was a more sensitive parameter than total organic C contents to identify changes in the soil due to land-use management.

Effects of Plant Residue Quality on C Dynamics

Vegetation can influence soil organic C levels as a result of the amount, placement and biodegradability of plant residues returned to the soil (Baldock and Nelson, 1999). The quantity and quality of added plant residues are important factors controlling C dynamics in different ecosystems, since it may regulate soil microbial biomass, affect C mineralization and organic matter dynamics (Dinesh et al., 2004). Residue mineralization rates depend on the quality of plant residues, environmental temperature, precipitation, and soil characteristics such as clay mineralogy, acidity, biological activity, and nutrient availability (Thönnissen et al., 2000). When environmental conditions are not limiting, the chemical and biochemical composition of plant residues are the main factors affecting decomposition (Trinsoutrot et al., 2000). Therefore, several works have recommended plant residue management as an efficient and sustainable land-use practice to increase biodiversity and biological process, efficiency of the nutrient cycling process, and soil organic matter (Matos et al., 2011a; Dersch and Böhm, 2001; Mueller and Koegel-Knabner, 2009). The management of different plant species with distinct capacity to remove or extract nutrients from the soil is especially important on agricultural systems with low use of fertilizer input. The high weathered soils of tropical regions due to low natural fertility, also present low content of SOM or soil chemical properties that are not suitable enough for plant growth.

Leguminous plants are considered to have a high potential for agroforestry systems intercropping in several regions, such as tropical Africa (Tian and Kang, 1998; Wortmann et al., 2000; Horst et al., 2001; Germani and Plenchette, 2004), Latin America (Barahona and Jansen, 2004; Meda and Furlani, 2005) and India (Bahl and Pasricha, 1998). However, the ability of leguminous plants to improve SOM quality and nutrient cycling are dependent on soil type, management and climate. These factors change the residue quality, usually defined in relation to its chemical and biochemical composition (Thönnissen et al., 2000). The C, N, P, lignin and polyphenols contents have been reported as good indicators of residue quality, since they are the most important regulators of organic material decomposition (Thomas and Asakawa, 1993; Mendonça and Stott, 2003). Working in a controlled environment, several researchers have found correlations between residue mineralization with its chemical and biochemical composition (Mafongoya et al., 2000; Mendonça and Stott, 2003), suggesting that this composition is the main controller of the decomposition rate.

However, the management of plant residues alone may be not enough to maintain C stocks in the soil under different land-use systems, especially in soils under intensive management systems. In a review, Sagrilo et al. (2011) concluded that leguminous plants used as green manures associated with appropriate soil management practices is a feasible and sustainable way to increase long-term SOM. They attribute this effect to the high amounts and quality of residues added to the soil, but also to the improvement of physical protection of SOM under conservation tillage practices.

Soil physical structure is closely related to soil C stocks. When forests are converted to arable lands an effective decrease in soil C stocks is expected (Schroth et al., 2002; Haile et al., 2008; Apezteguía et al., 2009), which can be attributed to reduced inputs of plant residues, increase of decomposability of crop residues, but mainly to the tillage effects that break-up aggregates and exposes organo-mineral surfaces (Post and Kwon, 2000), mainly to microbial community. In a review, Murty et al. (2002) assessed the changes in soil C caused by conversion of forest to cultivated land and observed that conversion led to an average of soil C loss of 30%. According to the same authors, carbon losses in soil converted from forest to cultivated land are associated to changes in soil physical, chemical and biological properties due to changes in the quantity and quality of crop residues inputs, nutrients inputs and losses, and stimulation of decomposition through soil disturbance.

Carbon Isotopic Composition as a Tool to Assess Soil Organic Matter Dynamics

Soil organic matter dynamics has also been investigated using the C isotopic composition technique in different ecosystems (Roscoe et al., 2001; Diels et al., 2004; Richards et al., 2007; Nii-Annang et al., 2009). Natural variations of C stable isotopes in the environment are used to trace the origins of organic matter pathways and the dynamics of its transformation (Balesdent and Mariotti, 1996; Bernoux et al., 1998; Roscoe et al., 2001; Diels et al., 2004). Most of these studies have been limited to evaluations of SOM dynamics based on natural differences in $\delta^{13}C$ isotopic signature where vegetation has changed from plants with the C3 photosynthetic pathway to C4 pathway or vice versa (Bernoux et al., 1998; Roscoe et al., 2001; Blagodatskaya et al., 2011).

However, there are many others possibilities to evaluate SOM dynamics using the $\delta^{13}C$ isotopic composition technique. Fractions of SOM associated with clay and silt originate from different positions in the soil with different exposures to microbial decomposition, which results in significant differences in chemical composition and in ^{13}C signature (Christensen, 2001). The evaluation of $\delta^{13}C$ could indicate the age and degree decay of the C associated with SOM and its fractions. Thus, changes in the $\delta^{13}C$ values of these fractions might reflect SOM turnover rate, and provide important information concerning the effects of different management system in the global C cycle (Bernoux et al., 1998). The ^{13}C technique in combination with modeling could also be used to analyze SOM changes in long and short-term experiments under different management systems (Diels et al., 2004). The key point is that it requires information on the quantity and ^{13}C natural abundance of each plant component of the system.

The application of ^{13}C stable isotope technique has been used to better understand the soil dynamics of C in agroforestry elsewhere (Lehmann et al., 1998; Diels et al.,

2001; Oelbermann et al., 2006b; Ares et al., 2009). Matos et al. (2011b) used the $\delta^{13}C$ technique to understand organic matter dynamics in soil under a silvopastoral system over the past 50 years in northeastern Germany and observed increase of $\delta^{13}C$ values of SOM fractions with increasing depth. They also observed $\delta^{13}C$ enrichment in the SOM fraction, which is associated with clay and silt. These finding were probably related to discrimination of ^{13}C relative to ^{12}C during organic matter decomposition, which may result in ^{13}C enrichment of the remaining SOM (Accoe et al., 2002). Evaluating the effect of tropical agroforestry systems on C sequestration, Takimoto et al. (2009) used ^{13}C isotopic ratio to assess the contribution of trees (C3 plants) and crops (C4 plants) to total soil C and observed that, more C from tree was found in larger particle size fraction and surface soil. They also observed that long-term tree presence contributed to enhance C sequestration in deeper soil layers.

Diels et al. (2001) working with long-term agroforestry system under a sub-humid tropical climate observed that annual variations in plant $\delta^{13}C$ values were of the order of only 0.4–0.6%, which correspond to only 3–4% of the maximum tracer signal difference between C3 and C4 plants. These results indicate that plant sampling and isotopic analysis can be limited to two-three years or even a single representative year in terms of weather. Quantifying C and N pools, gross SOC turnover and residue stabilization efficiency in the alley crop systems an American tropical region, Oelbermann et al. (2006b) showed the $\delta^{13}C$ signature of the soil shifted significantly towards that of C3 vegetation in the alley crop due to the greater input of organic residues from tree prunings compared to the sole crop. The proportion of input from tree prunings only in the 19-year-old alley crop ranged from 14 to 20%, and from 9 to 11% in the 10-year-old system to a soil depth of 20 cm. However, pruning may decrease the total root length density of sole cropped trees by 47% (Lehmann et al., 1998). The same authors, showed soil water depletion is higher under the tree row than in the alley and higher in alley cropping than in monocultural systems. Water competition between tree and crop may be confirmed by the carbohydrate analyses showing lower sugar contents of roots in agroforestry than in monoculture. The agroforestry combination used the soil water between the hedgerows more efficiently than the sole cropped trees or annual crops, as water uptake of the trees reached deeper and started earlier than the crops.

Final Comments

It is well known that soil conservation practices are a key factor in maintaining and increasing soil C stocks. But, in reality many agricultural lands around the world do not have this, facing huge losses of C due to aggregates breakdown and, consequently, soil erosion. Some governments have policies to control this and to push farmers to work with more soil friendly practices. One of these political initiatives is related to establishing agricultural programs like low C agriculture systems. In these programs, farmers are encouraged to adopt mechanical (as building terraces and so on), and edaphic (as applying fertilizers, green manure, vegetal soil cover, mulch and so on) soil conservation practices, in association to no-till, agroforestry, silvopastoral system, among others.

In the tropics, where farmers struggle to keep soil production in low input systems, agroforestry systems bring great opportunities to maintain and, also, improve soil quality together with high food production. However, the adoption of agroforestry systems may be achieved if future research finds solutions to solve the farmer's daily challenges like suitable tree species and appropriated tree management, which in combination with annual or perennial crops have low influence on agricultural crops. Moreover, further research should always take into account the potential benefits of agroforestry systems for a sustainable agriculture. The search for sustainable and productive systems can run through the proper management of the available resources while satisfying human needs, maintaining or improving environmental quality and conserving natural resources.

References

Accoe, F., Boeckx, P., Van Cleemput, O., Hofman, G., Zhang, Y., Li, R. and Guanxiong, C. 2002. Evolution of the $\delta^{13}C$ signature related to total carbon contents and carbon decomposition rate constants in a soil profile under grassland. Rapid Commun. Mass Spectrom. 16: 2184–2189.

Aguiar, A.C.F., Cândido, C.S., Carvalho, C.S., Monroe, P.H.M. and Moura, E.G. 2013. Organic matter fraction and pools of phosphorus as indicator of the impact of land use in the Amazonian periphery. Ecol. Indic. 30: 158–164.

Albrecht, A. and Kandji, S.T. 2003. Carbon sequestration in tropical agroforestry systems. Agric. Ecosyst. Environ. 99: 15–27.

Apezteguía, H.P., Izaurralde, R.C. and Sereno, R. 2009. Simulation study of soil organic matter dynamics as affected by land use and agricultural practices in semiarid Córdoba, Argentina. Soil Till. Res. 102: 101–108.

Ares, A., Burner, D.M. and Brauer, D.K. 2009. Soil phosphorus and water effects on growth, nutrient and carbohydrate concentrations, $\delta^{13}C$, and nodulation of mimosa (*Albizia julibrissin* Durz.) on a highly weathered soil. Agrofor. Syst. 76: 317–325.

Baldock, J.A. and Nelson, P.N. 1999. Soil organic matter. pp. B25–B84. *In*: Sumner, M. (ed.). Handbook of Soil Science. CRC Press, Boca Raton, FL.

Bahl, G.S. and Pasricha, N.S. 1998. Efficiency of P utilization by pigeonpea and wheat grown in a rotation. Nutri. Cycl. Agroecosyst. 51: 225–229.

Balesdent, J. and Mariotti, A. 1996. Measurement of soil organic matter turnover using ^{13}C natural abundance. pp. 83–111. *In*: Boutton, T.W. and Yamasaki, S. (eds.). Mass Spectrometry of Soils. Marcel Dekker, New York.

Barahona, M. and Janssen, M. 2004. Survey of green manure/cover crop systems of smallholder farmers in the tropics. pp. 286–301. *In*: Eilittä, M., Mureithi, J. and Derpsch, R. (eds.). Green Manure/Cover Crop Systems of Smallholder Farmers: Experiences from Tropical and Subtropical Regions. Springer, Netherlands.

Bayer, L.B., Batjes, N.H. and Bindraban, P.S. 2010. Changes in organic carbon stocks upon land use conversion in the Brazilian Cerrado: a review. Agric. Ecosyst. Environ. 137: 47–58.

Bernoux, M., Cerri, C.C., Neill, C. and de Moraes, J.F.L. 1998. The use of stable carbon isotopes for estimating soil organic matter turnover rates. Geoderma 82: 43–58.

Blagodatskaya, E., Yuyukina, T., Blagodatsky, S. and Kuzyakov, Y. 2011. Turnover of soil organic matter and microbial biomass under C3-C4 vegetation change: consideration of ^{13}C fractionation and preferential substrate utilization. Soil Biol. Biochem. 43: 159–166.

Bungart, R. and Hüttl, R.F. 2004. Growth dynamics and biomass accumulation of 8-year-old hybrid poplar clones in a short-rotation plantation on a clayey-sandy mining substrate with respect to plant nutrition and water budget. Eur. J. For. Res. 123: 105–115.

Cardoso, I.M., Van der Meer, P., Oenema, O., Janssen, B.H. and Kuyper, T.W. 2003. Analysis of phosphorus by 31PNMR in Oxisols under agroforestry and conventional coffee systems in Brazil. Geoderma 112: 51–70.

Chabbi, A. and Rumpel, C. 2009. Organic matter dynamics in agro-ecosystems—the knowledge gaps. Eur. J. Soil Sci. 60: 153–157.

Christensen, B.T. 2001. Physical fractionation of soil and structural and functional complexity in organic matter turnover. Eur. J. Soil Sci. 52: 345–353.

Dagang, A. and Nair, P.K.R. 2003. Silvopastoral research and adoption in Central America: recent findings and recommendations for future directions. Agrofor. Syst. 59: 149–155.

Dersch, G. and Böhm, K. 2001. Effects of agronomic practices on the soil carbon storage potential in arable farming in Austria. Nutri. Cycl. Agroecosyst. 60: 49–55.

Diels, J., Vanlauwe, B., Sanginga, N., Colen, E. and Merckx, R. 2001. Temporal variations in plant $\delta^{13}C$ values and implications for using the ^{13}C technique in long-term soil organic matter studies. Soil Biol. Biochem. 33: 1245–1251.

Diels, J., Vanlauwe, B., Van der Meersch, M.K., Sanginga, N. and Merckx, R. 2004. Long-term soil organic carbon dynamics in a subhumid tropical climate: ^{13}C data in mixed C_3/C_4 cropping and modeling with ROTHC. Soil Biol. Biochem. 36: 1739–1750.

Dinesh, R., Suryanarayana, M.A., Ghoshal Chaudhuri, S. and Sheeja, T.E. 2004. Long-term influence of leguminous cover crops on the biochemical properties of a sandy clay loam Fluventic Sulfaquent in a humid tropical region of India. Soil Till. Res. 77: 69–77.

Dixon, R. 1995. Agroforestry systems: sources of sinks of greenhouse gases? Agrofor. Syst. 31: 99–116.

Dube, F., Couto, L., Silva, M.L., Leite, H.G., Garcia, R. and Araujo, G.A.A. 2002. A simulation model for evaluating technical and economic aspects of an industrial eucalyptus-based agroforestry system in Minas Gerais, Brazil. Agrofor. Syst. 55: 73–80.

Freitas, P.L., Blancaneaux, P., Gavinelli, E., Larré-Larrouy, M. and Feller, C. 2000. Nível e natureza do estoque orgânico de Latossolos sob diferentes sistemas de uso e manejo. *Revista Pesquisa Agrpecuária Brasileira*, Pesq. Agropec. Bras. 35: 157–170.

Germani, G. and Plenchette, C. 2004. Potential of *Crotalaria* species as green manure crops for the management of pathogenic nematodes and beneficial mycorrhizal fungi. Plant Soil 266: 333–342.

Gruenewald, H., Brandt, B.K.V., Schneider, B.U., Bens, O., Kendzia, G. and Hüttl, R.F. 2007. Agroforestry systems for the production of woody biomass for energy transformation purposes. Ecol. Eng. 29: 319–328.

Haile, S.G., Nair, P.K.R. and Nair, V.D. 2008. Carbon storage of different soil-size fractions in Florida silvopastoral systems. J. Environ. Qual. 37: 1789–1797.

Horst, W.J., Kamh, M., Jibrin, J.M. and Chude, V.O. 2001. Agronomic measures for increasing P availability to crops. Plant Soil 237: 211–223.

Jose, S., Gillespie, A.R. and Pallardy, S.G. 2004. Interspecific interactions in temperate agroforestry. Agrofor. Syst. 61-62: 237–255.

Kang, B.T. 1997. Alley cropping—soil productivity and nutrient recycling. For. Ecol. Manage. 91: 75–82.

Leblanc, H.A., Nygren, P. and McGraw, R.L. 2006. Green mulch decomposition and nitrogen release from leaves of two *Inga* spp. in an organic alley-cropping practice in the humid tropics. Soil Biol. Biochem. 38: 349–358.

Lehmann, J., Peter, I., Steglich, C., Gebauer, G., Huwe, B. and Zech, W. 1998. Below-ground interactions in dryland agroforestry. For. Ecol. Manage. 111: 157–169.

Leifeld, J. and Kögel-Knabner, I. 2005. Soil organic matter fractions as early indicators for carbon stock changes under different land-use? Geoderma 124: 143–155.

Leite, L.F.C., Mendonça, E.S., Machado, P.L.O.A. and Matos, E.S. 2003. Total C and N storage and organic C pools of a Red-Yellow Podzolic under conventional and no tillage at the Atlantic Forest Zone, south-eastern Brazil. Aust. J. Soil Res. 41: 717–730.

Maia, S., Xavier, F., Oliveira, T., Mendonça, E. and Araújo Filho, J. 2007. Organic carbon pools in a Luvisol under agroforestry and conventional farming systems in the semi-arid region of Ceará, Brazil. Agrofor. Syst. 71: 127–138.

Mafongoya, P.L., Barak, P. and Reed, J.D. 2000. Carbon, nitrogen and phosphorus mineralization of tree leaves and manure. Biol. Fertil. Soils 30: 298–305.

Malézieux, E., Crozat, Y., Dupraz, C., Laurans, M., Makowski, D., Ozier-Lafontaine, H., Rapidel, B., de Tourdonnet, S. and Valantin-Morison, M. 2009. Mixing plant species in cropping systems: concepts, tools and models. A review. Agron. Sustain. Dev. 29: 43–62.

Matos, E.S., Cardoso, I.M., Souto, R.L., Lima, P.C. and Mendonça, E.S. 2011a. Characteristics, residue decomposition and carbon mineralization of leguminous and spontaneous plants in coffee systems. Commun. Soil Sci. Plant Anal. 42: 489–502.

Matos, E.S., Freese, D., Mendonça, E.S., Slazak, A. and Hüttl, R.F. 2011b. Carbon, nitrogen and organic C fractions in topsoil affected by conversion from silvopastoral land use systems. Agrofor. Syst. 81: 203–211.

Meda, A.R. and Furlani, P.R. 2005. Tolerance to aluminium toxicity by tropical leguminous plants used as cover crops. Braz. Arch. Biol. Techn. 48: 309–317.

Mendonça, E.S. and Stott, D.E. 2003. Characteristics and decomposition rates of pruning residues from a shaded coffee system in Southeastern Brazil. Agrofor. Syst. 57: 117–125.

Michel, G.A., Nair, V. and Nair, P. 2007. Silvopasture for reducing phosphorus loss from subtropical sandy soils. Plant Soil 297: 267–276.

Montagnini, F. and Nair, P.K.R. 2004. Carbon sequestration: an underexploited environmental benefit of agroforestry systems. Agrofor. Syst. 61-62: 281–295.

Moura, E.G., Serpa, S.S., Santos, J.G.D., Sobrinho, J.R.C. and Aguiar, A.C.F. 2010. Nutrient use efficiency in alley cropping systems in the Amazonian periphery. Plant Soil 335: 363–371.

Mueller, C. and Koegel-Knabner, I. 2009. Soil organic carbon stocks, distribution, and composition affected by historic land use changes on adjacent sites.Biol. Fertil. Soils 45: 347–359.

Muñoz, C., Zagal, E. and Ovalle, C. 2007. Influence of trees on soil organic matter in Mediterranean agroforestry systems: an example from the 'Espinal' of central Chile. Eur. J. Soil Sci. 58: 728–735.

Murage, E.W., Voroney, P. and Beyaert, R.P. 2007. Turnover of carbon in the free light fraction with and without charcoal as determined using the ^{13}C natural abundance method. Geoderma 138: 133–143.

Murty, D., Kirschbaum, M.U.F., McMurtrie, R.E. and McGilvray, H. 2002. Does conversion forest to agricultural land change soil carbon and nitrogen? A review of the literature. Glob. Change Biol. 8: 105–123.

Mutuo, P.K., Cadisch, G., Albrecht, A., Palm, C.A. and Verchot, L. 2005. Potential of agroforestry for carbon sequestration and mitigation of greenhouse gas emissions from soils in the tropics. Nutri. Cycl. Agroecosyst. 71: 43–54.

Nair, P.K.R. 1985. Classification of agroforestry systems. Agrofor. Syst. 3: 97–128.

Nair, V.D., Haile, S.G., Michel, G.-A. and Nair, P.K.R. 2007. Environmental quality improvement of agricultural lands through silvopasture in southeastern United States. Scientia Agricola 64: 513–519.

Nair, P.K.R., Kumar, M. and Nair, V.D. 2009. Agroforestry as a strategy for carbon sequestration. J. Plant Nutr. Soil Sci. 172: 10–23.

Neff, J.C., Townsend, A.R., Gleixner, G., Lehman, S.J., Turnbull, J. and Bowman, W.D. 2002. Variable effects of nitrogen additions on the stability and turnover of soil carbon. Nature 419: 915–917.

Nii-Annang, S., Grünewald, H., Freese, D., Hüttl, R. and Dilly, O. 2009. Microbial activity, organic C accumulation and ^{13}C abundance in soils under alley cropping systems after 9 years of recultivation of quaternary deposits. Biol. Fertil. Soils 45: 531–538.

Oelbermann, M., Voroney, R.P., Kass, D.C.L. and Schlönvoigt, A.M. 2005. Above- and below-ground carbon inputs in 19-, 10- and 4-year-old Costa Rican alley cropping systems. Agric. Ecosyst. Environ. 105: 163–172.

Oelbermann, M., Voroney, R., Thevathasan, N., Gordon, A., Kass, D. and Schlönvoigt, A. 2006a. Soil carbon dynamics and residue stabilization in a Costa Rican and southern Canadian alley cropping system. Agrofor. Syst. 68: 27–36.

Oelbermann, M., Voroney, R.P., Kass, D.C.L. and Schlönvoigt, A.M. 2006b. Soil carbon and nitrogen dynamics using stable isotopes in 19- and 10-year-old tropical agroforestry systems. Geoderma 130: 356–367.

Oelbermann, M. and Voroney, R.P. 2007. Carbon and nitrogen in a temperate agroforestry system: using stable isotopes as a tool to understand soil dynamics. Ecol. Eng. 29: 342–349.

Pacheco, A.R., Chaves, R.Q. and Nicoli, C.M.L. 2013. Integration of crops, livestock, and forestry: a system of production for the Brazilian Cerrados. pp. 51–61. *In*: Hershey, C. and Neate, P. (eds.). Eco-Efficiency: From Vision to Reality. CIAT, Cali.

Palm, C., Blanco-Canqui, H., DeClerk, F., Gatere, L. and Grace, P. 2014. Conservation agriculture and ecosystem services: an overview. Agric. Ecosyst. Environ. 187: 87–105.

Post, W.M. and Kwon, K.C. 2000. Soil carbon sequestration and land-use change: processes and potential. Glob. Change Biol. 6: 317–327.

Reynolds, P.E., Simpson, J.A., Thevathasan, N.V. and Gordon, A.M. 2007. Effects of tree competition on corn and soybean photosynthesis, growth, and yield in a temperate tree-based agroforestry intercropping system in southern Ontario, Canada. Ecol. Eng. 29: 362–371.

Richards, A.E., Dalal, R.C. and Schmidt, S. 2007. Soil carbon turnover and sequestration in native subtropical tree plantations. Soil Biol. Biochem. 39: 2078–2090.

Roscoe, R., Buurman, P., Velthorst, E.J. and Vasconcellos, C.A. 2001. Soil organic matter dynamics in density and particle size fractions as revealed by the 13C/12C isotopic ratio in a Cerrado's oxisol. Geoderma 104: 185–202.

Sagrilo, E., Leite, L.F.C. and Maciel, G.A. 2011. Soil organic matter as affected by green manure at Brazilian conditions. Dyn. Soil Dyn. Plant. 5: 7–11.

Schimel, D.S. 1995. Terrestrial ecosystems and the carbon cycle. Global Change Biology 1: 77–91.

Schroth, G., D'Angelo, S.A., Teixeira, W.G., Haag, D. and Lieberei, R. 2002. Conversion of secondary forest into agroforestry and monoculture plantations in Amazonia: consequences for biomass, litter and soil carbon stocks after 7 years. For. Ecol. Manage. 163: 131–150.

Sohi, S.P., Mahieu, N., Arah, J.R.M., Powlson, D.S., Madari, B. and Gaunt, J.L. 2001. A procedure for isolating soil organic matter fractions suitable for modeling. Soil Sci. Soc. Am. J. 65: 1121–1128.

Tapia-Coral, S.C., Luizão, F.J., Wandelli, E. and Fernandes, E.C.M. 2005. Carbon and nutrient stocks in the litter layer of agroforestry systems in central Amazonia, Brazil. Agroforestry Systems 65: 33–42.

Takimoto, A., Nair, V.D. and Nair, P.K.R. 2009. Contribution of trees to soil carbon sequestration under agroforestry systems in the West African Sahel. Agrofor. Syst. 76: 11–25.

Thomas, R.J. and Asakawa, N.M. 1993. Decomposition of leaf litter from tropical forage grasses and legumes. Soil Biol. Biochem. 25: 1351–1361.

Thönnissen, C., Midmore, D.J., Ladha, J.K., Olk, D.C. and Schmidhalter, U. 2000. Legume decomposition and nitrogen release when applied as green manures to tropical vegetable production systems. Agron. J. 92: 253–260.

Tian, G. and Kang, B.T. 1998. Effects of soil fertility and fertilizer application on biomass and chemical compositions of leguminous cover crops. Nutri. Cycl. Agroecosyst. 51: 231–238.

Trinsoutrot, I., Recous, S., Bentz, B., Lineres, M., Cheneby, D. and Nicolardot, B. 2000. Biochemical quality of crop residues and carbon and nitrogen mineralization kinetics under nonlimiting nitrogen conditions. Soil Sci. Soc. Am. J. 64: 918–926.

United States. 2013. Department of Agriculture Natural Resources Conservation Service: Soil Health. Available in: <http://www.nrcs.usda.gov/wps/portal/nrcs/main/soils/health/> Acessed 26 August 2013.

Unite States. 2014. United States Environmental Protection Agency. Carbon Dioxide Capture and Sequestration. Available in: <http://www.epa.gov/climatechange/ccs/index.html> Accessed 14 February 2014.

Wortmann, C.S., McIntyre, B.D. and Kaizzi, C.K. 2000. Annual soil improving legumes: agronomic effectiveness, nutrient uptake, nitrogen and water use. Field Crops Res. 68: 75–83.

Wu, T., Schoenau, J.J., Li, F., Qian, P., Malhi, S.S. and Shi, Y. 2003. Effect of tillage and rotation on organic carbon forms of chernozemic soils in Saskatchewan. J. Plant Nutr. Soil Sci. 166: 328–335.

7

Forests, Land Use Change, and Water

Devendra M. Amatya,[1,*] *Ge Sun,*[2] *Cole Green Rossi,*[3]
Herbert S. Ssegane,[4] *Jami E. Nettles*[5] and *Sudhanshu Panda*[6]

Introduction

A forest is a biotic community predominated by trees and woody vegetation that are significantly taller, greater, thicker, and deeper than other vegetation types and generally covers a large area (Chang, 2003). Forests cover approximately 26.2% of the world, with 45.7% of Latin America and the Caribbean being covered, 35% of East Asia and the Pacific, and 35% of the European Union. Canada and the United States (U.S.) combined account only for 6.8% of the world's forests while Africa has even less 5.7% (Forest Types of the World, 2013). In the U.S., forests cover about one-third of its land (Sedell et al., 2000; Jones et al., 2009), totaling about 300 million ha (USDA, 2001). Forested areas in the temperate zone have not changed much in

[1] USDA Forest Service, Center for Forested Wetlands Research, 3734 Highway 402, Cordesville, SC 29434, USA.
Email: damatya@fs.fed.us
[2] USDA Forest Service, Eastern Forest Environmental Threat Assessment Center, 920 Main Campus Dr., Venture II, Suite 300, Raleigh, NC 27606, USA.
Email: gesun@fs.fed.us
[3] USDOI Bureau of Land Management, 440 West, 200 South, Salt lake City, UT 84101.
Email: crossi@blm.gov
[4] Argonne National Lab, Energy Systems Division, Lemont, IL 60439.
Email: hssegane@anl.gov
[5] Weyerhaeuser Company, P.O. Box 2288, Columbus, MS.
Email: Jami.Nettles@weyerhaeuser.com
[6] University of North Georgia, 3820 Mundy Mill Road, Oakwood, GA 30566.
Email: Sudhanshu.Panda@ung.edu
* Corresponding author

recent decades, but continuing deforestation of tropical forests, about half of world total, is of great concern (World Resources Institute, 1996).

Land use and forests are intricately linked to how and where people live and sustain themselves (GEF, 2012). The livelihood of more than one billion people depends on tropical forests (Lynch et al., 2013). Similarly, water is critical for human life, for many human activities as well as an environmental resource (EOS, 2012). Worldwide, early human society and culture is tied to trees, forests, and water. The connections between the loss of forests, land use, streamflow, and water quality have long been recognized (de la Cretaz and Barten, 2007). An example of prehistoric Athens, was used by these authors, where the originally wooded lands were left with bare dry soil just like skin and bone resulting in flooding and dried springs due to the cutting of these forests. As a result of large-scale expansion of croplands and pasturelands at the expense of forests and grasslands, cultivated cropland and pastureland have increased globally by 460% and 560%, respectively in the past 300 years (Scanlon et al., 2007).

The major causes of land use change (LUC) since 1750 has been deforestation of temperate regions for food production and industrialization. Modern LUCs from forests to a landscape with a mosaic of agricultural, forest and urban lands have resulted in new environmental issues including landslides, flooding, soil erosion, water quantity and quality degradation, salinization, desertification, and ecosystem service losses (Amatya et al., 2009; Sun and Lockaby, 2012).

Foresters started to work proactively to better understand the relationships of forests and water in the early 20th century (Chang, 2003; Andreassian, 2004). Water and forests are recognized as two important resources that provide vital habitat for wildlife, clean air and water (Brown et al., 2008; Jones et al., 2009; FAO, 2013), recreation, timber (Prestemon and Abt, 2002), and bioenergy (King et al., 2013). Most importantly, forests are large carbon sinks and play an increasingly important role in mitigating global climate change (Bonan, 2008; Dai et al., 2013). Forest and water are interdependent natural resources; the connection between water and forests is recognized with the birth of professional forestry in the U.S. (Ice and Stednick, 2004). As a result, forest experimental watershed studies that were designed to understand forest hydrologic processes and answer forest-water relations were initiated in federal lands in the beginning of 20th century with the first one as the Wagon Wheel Gap forest in Oregon (Bates, 1921) and forest hydrologic research continues till today to refine our understanding of the water cycle in forests (Chang, 2003; Jones et al., 2009; Vose et al., 2011). Land use has changed rapidly in several parts of the world in the last few decades as a part of the global change phenomena (De Fries and Eshelman, 2004; Scanlon et al., 2007; GEF, 2012), and especially true in the U.S. (Clifton et al., 2006; Hamilton et al., 2008; U.S. Forest Service, 2011; Sun and Lockaby, 2012). Land use change may occur due to change in vegetation such as deforestation, afforestation, urbanization, and other kinds of land development including mining and construction of highways. Accordingly, there have been increased concerns about the impacts of LUC on flooding, streamflow (yields), baseflow, and quality of waters draining from the uplands into downstream water bodies. Hydrologic impacts of forest conversions are critical to issues of contaminant dilution, aquatic habitat, and public water supply and use (Wilk et al., 2001; Tang et al., 2005; Thanapakpawin et al., 2006; Clifton et al., 2006; Skaggs et al., 2011; Price et al., 2011; Vose et al., 2012). The need to increase agricultural production to feed a growing world population leads to even more concerns about environmental impacts of converting forest and pasture lands to row crop agriculture (Skaggs et al., 2011). Emphasis on growing energy crops for biofuel production will potentially increase conversion of forests and other lands to intensively cultivated fields (King et al., 2013).

In its first comprehensive forecast on southern forests, the U.S. Forest Service (2011) stated that urbanization, bioenergy use, weather patterns, land ownership changes, and invasive species will significantly alter the South's forests between the years 2010 and 2060. The area of forest land is projected to decrease by about 9.3 million ha, mainly due to population growth and urbanization.

DeFries and Eshleman (2004) suggested a need for understanding the consequences of LUC for hydrologic processes, and integrating this understanding into the emerging focus on LUC science. Scanlon et al. (2007) provided a comprehensive review and summary on global impacts of conversions from natural to agricultural ecosystems on quantity and quality of water resources for both surface and groundwater, and addressed some of those consequences in water demand, supply, and water quality. There are several studies in the literature from around the world on impacts of forest clearing on downstream hydrology and water yield (Bosch and Hewlett, 1982; Andreassian, 2004; Brown et al., 2005). Most of these studies suggest that forest management practices such as harvesting, or the conversion of forests to agricultural or other uses increase in streamflows, water table levels, and increased groundwater recharge as a result of reduced evapotranspiration (ET) (Stednick, 1996; Sun et al., 2005; Amatya et al., 2006; Abdelnour et al., 2011; Skaggs et al., 2011; Webb et al., 2012; Tian et al., 2012). However, Farley et al. (2005) and Sun et al. (2006) noted that there is only a limited knowledge on a systematic analysis of the effects of afforestation (i.e., conversion of grass, shrub, or croplands to forests) on watershed hydrology.

The impact of LUCs on water resources also depends on many factors, including the original vegetation being replaced, the vegetation replacing it, the type of change, and associated land management and application practices (Scanlon et al., 2007), upon the dominant soil type where the LUC occurs. Local climate, extreme events, and soils are important factors to consider (Jayakaran et al., 2014; Boggs et al., 2012; Caldwell et al., 2012).

Efforts to determine the hydrologic impacts of LUC have been conducted on a wide range of scales using a relatively large range of methods (Skaggs et al., 2011). Methods vary from simply monitoring precipitation, streamflow, and other basic hydrologic variables like land use/land cover (LULC), elevation, slope, etc. of the watershed during and following land use conversion, to data intensive paired-watershed approaches, to the application of models ranging from simple regression methods to process-based integrated models. However, there is only limited synthesized information on these assessment methods including the change detection. Similarly, there are knowledge gaps in understanding the effects of various specific factors including the potential evapotranspiration (PET) that varies with reference vegetation (also called reference-ET (REF-ET)) and climate change and ultimately may affect the hydrology and water quality assessments for land use conversion. Land use and climate are two main factors directly influencing watershed hydrology, and separation of their effects is of great importance for land use planning and management (Li et al., 2009).

In this paper we start by giving a brief background on the status of forest hydrologic balance and then review the current literature on available methods, scaling issues, and detection limits used for evaluating the impacts of LUC. Furthermore, specifics on effects of LUC such as water use by forests and crops or ET, change in soil hydraulic properties after forest harvesting, artificial drainage, urbanization that alters land imperviousness, and climate change, including extreme events are also considered in this synthesis.

The specific objectives are to (1) synthesize information on monitoring and modeling approaches, change detection and statistical methods in various scales including remote sensing method; (2) synthesize information on hydrologic effects of various factors including land use conversion and climate change; and (3) provide recommendations on future research directions.

Forest Water Balance

Because forests make up a relatively large portion of many of our watersheds, it is important to understand their water balance components and their flow paths and distribution for both natural forests and silvicultural operations, while considering the contribution of other land uses. Main components of the forest hydrologic balance include precipitation as input and canopy interception, throughfall, stemflow, surface runoff, quick and interflows, transpiration, understory and soil/litter evaporation, deep seepage as outputs through various pathways (i.e., forest canopy, root system, litter and soil) and change in soil-water storage (Fig. 7.1). Evapotranspiration is the sum of water loss through the process of rainfall interception from the tree canopies, transpiration

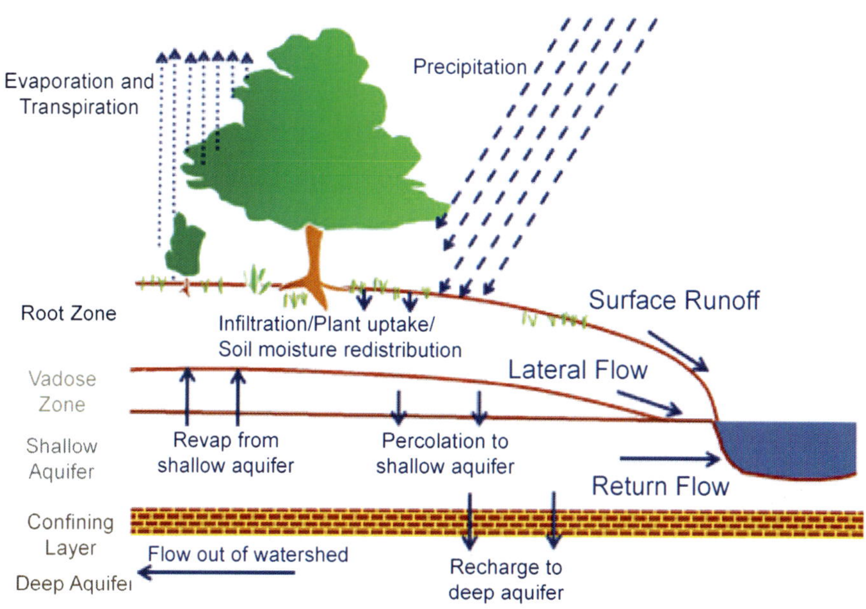

Figure 7.1. Schematic of forest hydrologic processes (Source: SWAT manual; Neitsch et al., 2005).

from foliage, and evaporation from forest floor. It is the key hydrologic flux that links water, energy, and biogeochemical cycles in forests (Sun et al., 2011a; 2011b).

Key forest hydrology questions identified by NRC (2008) for understanding basic processes and principles of water movement and predicting the general directions and magnitudes of hydrologic effects of anthropogenic and climate change were: (a) what are the flow paths and storage reservoirs of water in forests and forested watersheds? (b) how do modifications of forest vegetation influence water flowpaths and storage?, and (c) how do changes in forests affect water quantity and quality? de la Critaz and Barten (2007) provide a step-by-step approach on understanding hydrologic principles and processes mostly on forest landscape as a reference that govern the interactions between forest, water, and land use to experimental studies of varying scales and their management implications for the northeastern U.S. The authors also present the hydrologic and water quality principles to construct management plans for water supply watersheds on varying spatial and temporal scales.

The impact of changes in forest land use on its hydrology, in part, is reflected by the water balance components in equation 1:

$$\boldsymbol{P - E_i - E_{sl} - E_t - R_o - R_{gw} - DS = \Delta S} \qquad (1)$$

Where, $\boldsymbol{P}$ = Precipitation, $\boldsymbol{E_i}$ = Evaporation from canopy interception, $\boldsymbol{E_{sl}}$ = Evaporation from soil and litter, $\boldsymbol{E_t}$ = transpiration from over and understory, R_o = surface runoff, R_{gw} = subsurface quick and return flow (baseflow), and DS = Deep seepage. ET, ET is the sum of $\boldsymbol{E_i + E_{sl} + E_t}$. The sum of the components of $\boldsymbol{R_o - R_{gw}}$ is also total streamflow. On a longer (> 1 year) term basis $\boldsymbol{\Delta S}$ can be assumed negligible.

Surface runoff seldom occurs in forests with large surface depressional storage (Amatya and Skaggs, 2011; Amatya et al., 1996), thick litter layer, and high soil infiltration rates, and thus streamflow is derived mainly from subsurface flow and groundwater. Generally, ET is a major loss of water in forest hydrologic water balance (Brauman et al., 2012; Sun et al., 2011a; 2011b; Amatya and Skaggs, 2011; Tian et al., 2012; Zhang et al., 2012). Therefore, ET is important for water resources management and development, stream ecology and fluvial geomorphology (Sun et al., 2005; Zhang et al., 2001). The basic seasonal and annual forest water balance can be dramatically shifted depending on climatic conditions, vegetation types and dynamics, and soil type.

In contrast to our knowledge about the effects of forests on peakflow or floods, impacts of forest conversion on baseflow (low flow) at the large watershed scale are not clear and have not been well documented in literature. Recently, in their study of effects of watershed land use and geomorphology on stream low flows during severe drought conditions in the Blue Ridge mountains of the U.S., Price et al. (2011) found a consistent, significant positive relationship of watershed forest cover with low flows, despite the higher ET rates associated with forests compared with other land covers and despite the relatively small range of disturbance in the study area. New activities and forest products often emerge in response to specific needs. Growth in biofuel demand could lead to increased removal of biomass from plantation forests, resulting in substantial hydrologic impacts on these lands primarily due to reduction in ET.

Methods Used for Impact Assessment

Monitoring approach

Methods of monitoring usually involve measurements of temporal variables such as precipitation, streamflow (surface runoff), and weather parameters. Spatially distributed variables from watershed characteristics include LULC, slope, elevation, altitude, and land and soil management. These variables are used for a single watershed before and after treatment or land use conversion for a single watershed to a paired watershed approach. For example, Silvera and Alonso (2008) compared flow events from the 2100 km^2 Manuel Diaz basin in Uruguay before and after afforestation (25% of the watershed area). These authors also estimated decreases in annual streamflow between 8.2% to 36.5% after afforestation. Although a long time series of streamflow data from a single watershed can be used with change detection methods, such methods can mask the effects of annual and seasonal climatic variability. On the other hand, a paired watershed approach assumes that there is a consistent, quantifiable, and predictable relationship between watershed response variables (Ssegane et al., 2013).

Paired watershed approach

A large number of small field-scale experimental studies using a paired-watershed approach have been conducted in Australia, New Zealand, South Africa, South America, Great Britain, China, Japan, and the US to better understand forest hydrologic processes, their interactions with the environment, and their ecohydrologic impacts (Swank and Douglas, 1974; Bosch and Hewlett, 1982; Amatya et al., 1996; Sahin and Hall, 1996; Fahey and Jackson, 1997; Sun et al., 2001; Worrall et al., 2003; Andreassian, 2004; Jackson et al., 2004; Brown et al., 2005; Elliott and Vose, 2005; Farley et al., 2005; Amatya et al., 2006; Edwards and Troendle, 2008; Chescheir et al., 2009; Ssegane et al., 2013; Bren and Lane, 2014; Bren and Mcguire, 2012). More than 90 years after the first paired watershed study at Wagon Wheel Gap, Colorado (Bates, 1921), forest hydrologists and natural resources managers are still working to understand the effects and the variances of forest management practices on hydrology and water quality (Zegre, 2008).

The highly variable nature of watershed responses to disturbance, by harvesting, fire, insect and disease damage, or species replacement, depends on many factors, such as watershed scale, climate, forest and vegetation types, density, geology, soils, topography, elevation, aspect, disturbance location, and type of disturbance, and vegetation. Decades of field and experimental research have been conducted to evaluate the effects of disturbance on many watershed attributes, and in response several methods have been developed and employed. The paired watershed approach offers the ability to identify roles of forest cover, internal watershed behavior, and climate variability to establish a "baseline" for reference (Zegre, 2008). This approach continues to be used on low-order watersheds as the primary method for impact assessments (Bren and Lane, 2014); its validity for predicting effects on large flooding events had been challenged (Alila et al., 2009). Andreassian (2004) presented a summary of paired watershed results to help understand contradictions of the past, as well as highlight unresolved issues in forest hydrology. Examples of the paired-watershed approach using intensively monitored, relatively small watersheds include: Swank and Crossley (1988), Amatya et al. (2006), and Boggs et al. (2012) who studied effects of harvesting (clearcutting) on North Carolina mountain, drained coastal plain, and the piedmont landscapes, respectively.

Long-term watershed studies that integrate forests, land use or land cover change, and water use in Africa include five paired watershed studies in South Africa (Van Wyk, 1987; Smith and Scott, 1992; Scott and Lesch, 1997; Scott and Smith, 1997; Scott et al., 1998; Jewitt, 2002) at experimental watersheds of Cathedral Peak (Kwazulu-Natal province), Mokobulaan (Mpumalanga province), Westfalia (Limpopo province), and Jonkershoek (Western Cape province). The watersheds were established to quantify the effects of afforestation on streamflow. The control watersheds included grasslands at Cathedral Peak and Mokobulaan, and native scrub forests at Westfalia and Jonkershoek. Treatments included afforestation with *Eucalyptus grandis* at Westfalia and Mokobulaan, *pinus patula* at Mokobulaan and Cathedral Peak, and *pinus radiata* at Jonkershoek. Reductions in streamflow due to afforestation were a function of forest type (eucalyptus or pine), location of the watershed (optimal or suboptimal growth zone) and the number of years after afforestation. Total streamflow reductions responded faster under eucalyptus (100% reduction within 8 to 9 years) than pine trees (80 to 90% reduction within 16 to 22 years) due to a faster growth rate of eucalyptus. Also, although afforestation covered only 1.2% of the land cover in South Africa, it contributed 3.2% reduction in the total annual streamflow and 7.8% reduction of the low-flows. The low-flows were defined as flows in the driest three months of an average year based on a period of 70 years.

Dagg and Blackie (1965) describe paired watershed studies in Kenya and Tanzania (East Africa). The watershed sites were located in sub-humid climatic region where less than 4% of areal land cover received more than 1250 mm annual rainfall. At the Kericho site in Kenya, the control watershed was under Montane forest while the treatment was planted with tea (54% of the watershed). At Kimakia (Kenya), the control watershed was under Bamboo forest while the treatment was under softwood plantations. At Mbeya Range (Tanzania), the control watershed was under evergreen forest while the treatment was under locally cultivated crops. According to Edwards et al. (1976), the long-term average (1958–1973) water use over the above study watersheds decreased by 8.9% at the treatment watersheds compared to the control watersheds. For example, at Kericho the water use, calculated as the difference between rainfall and streamflow, decreased by 14.4% during clearing and planting (1960–1963), increased by 2.4% during tea plantation establishment (1964–1967), and decreased by 12.1% between 1968 and 1973. Additional reports and studies on the above watersheds include works by Blackie (1972) and Edwards and Blackie (1981).

In a more recent 8-year study on the conversion of grasslands to managed pine forest on smaller paired watersheds in Uruguay, Chescheir et al. (2009) found no reduction in the third year to a 28% reduction in the fourth year since tree planting.

The year with the greatest yield reduction was characterized by a very dry period followed by a very wet one. The water yield reduction over the last three years of the study was 15%.

Flow Duration Curve (FDC) as a visual tool for change detection

A review of paired watershed studies demonstrates the relevance of the flow duration curve (FDC) as a graphical tool to detect impacts of LUC on different flow regimes (high-flows, medium-flows, and low-flows) (Best, 2003). Best (2003) examined three watersheds: Redhill (Australia; pasture to pine forest), Wights (Australia; native forest to pasture), and Glendhu (New Zealand; grassland to pine). To minimize the effects of climatic variability, daily FDCs with similar annual rainfall of about 880 mm over eight years were compared at the Redhill watershed. The two years with similar annual rainfall coincided with the first and eighth years after pine planting. Comparison of FDCs at one and eight years after pine planting showed a 50% reduction in high-flows and 100% reduction in low-flows. Also, the observed increases in streamflow magnitudes at Wights watershed due to LUC (forest to pasture) were comparable to observed reductions at Redhill for the respective flow-regimes (high and low-flows). However, the conversion of grassland to pine plantation at the Glendhu watershed on average reduced the different flow-regimes by 30%. However, Lane et al. (2003) highlight the need to improve the understanding of the impact of afforestation on the FDC. These authors found their flow reductions were in accordance with published results for paired watershed studies but with two different patterns (one with more zero flows and another with a uniform reduction across all percentiles) for 10 watersheds they studied. They also suggested the usefulness of their model in removing the effect of rainfall variability, thus making it applicable where paired watershed data are not available.

Uncertainty of calibration data may mask small treatment effects

Laurén et al. (2009) demonstrate how uncertainty in pre-treatment data of paired watershed studies may influence estimates of the magnitude and duration of treatment effects. The monitoring of phosphorous loads on two independent paired watersheds in Finland before and after clear-cutting demonstrated that small treatment effects may be masked by uncertainty of the pre-treatment data. Bonumá et al. (2013) state that their model simulations could not capture the runoff peaks well in the daily flow record possibly due to uncertainty in the modified CN2 method used to estimate surface runoff (Mishra and Singh, 2003). In the case where the time of concentration of the watershed is less than 1 day, the uncertainty in estimated surface runoff from daily rainfall is greater. Green et al. (2006) argue that as one value represents the range of rainfall intensities that can occur within a day, there can be a considerable uncertainty within that time period that are not captured.

The understanding of basic hydrologic processes and their interactions gained in paired watershed and other experimental studies has enabled the development of more reliable simulation models (Skaggs et al., 2011) that can capture the small treatment effects. Most recently, Andreassian et al. (2012) demonstrated how a classical hydrologic model and a paired watershed model can be associated to reach an unprecedented level of efficiency. The authors reported that such a combined method can be useful for hydrological applications including trend analysis (i.e., streamflow after LUC).

Alila et al. (2009) demonstrate how an inappropriate pairing of floods by meteorological input in analysis of covariance (ANCOVA) and analysis of variance (ANOVA) statistical tests used extensively for evaluating the effects of forest harvesting on floods smaller and larger than an average event, leads to incorrect estimates of changes in flood magnitude because neither the tests nor the pairing, account for changes in flood frequency. Similarly, Kuras et al. (2012) argued that contrary to the prevailing perception in forest hydrology, the effects of harvesting are found to increase with return period. This result is attributable to the uniqueness of peak flow runoff generation processes in snow-dominated watersheds.

Hydrologic modeling

Detecting hydrologic effects of land conversions using the 'paired watershed' approach can be time consuming, expensive and cost-prohibitive, and often limited by treatment options (i.e., watershed location and size and vegetation manipulation types) and understanding interaction of LUC and climatic variability. So, hydrological models have been widely used in such investigations and model simulation studies are frequently conducted to assess the impacts of LUC on large basins (Lorup et al., 1998; Wilk et al., 2001; Siriwardena et al., 2006; Gassman et al., 2007; Breuer et al., 2009; Simin et al., 2011).

Hydrologic models vary from lumped to physically-based, distributed watershed-scale for assessing the hydrologic impacts of LUC (Singh and Frevert, 2006). Breuer et al. (2009) examined a set of 10 lumped, semi-lumped and fully distributed hydrologic models that have been previously used in LUC studies in low mountain watersheds of Germany. The authors found a substantial difference in model performance that was attributed to model input data, calibration, and the physical basis of the models. The effect of the physical differences between models on the long-term water balance was mainly attributed to differences in how models represent ET. The authors concluded that there was no superior model if several measures of model performance were considered and that all models were suitable to participate in further multi-model ensemble set-ups and LUC scenario simulations. In a companion study with a scenario analysis, Huisman et al. (2009) reported that there was a 90% general agreement about the direction of changes in the mean annual discharge by the ensemble members.

Application of these models has been greatly enhanced by the development of GIS-based data sources for soils, stream locations and characteristics, and the type and distribution of vegetation via satellite data. A common approach is to use observed hydrologic data to calibrate a simulation model for the current land use, followed by prediction of outflow and other hydrologic variables for conditions after conversion. If a paired-watershed study is conducted, the model can be calibrated and tested for both land uses. This procedure was followed by von Stackelberg et al. (2007), who used the SWAT model to determine that afforestation of a pastured watershed in northern

Uruguay would decrease average annual outflow (yield) by 23%. A potential source of error in this approach is that model inputs, such as hydraulic properties of the soil, may be affected by the change in land use and not properly reflected in the predictions (Heuvelmans et al., 2004). This was found to be the case for hydraulic conductivity of the upper 90 cm of the soil profile as evidenced by Skaggs et al. (2006) and will be investigated herein.

Simulation results obtained by Kim et al. (2012) revealed that increased mean annual outflow was significant ($\alpha = 0.05$) for 100% conversion from forest (261 mm) to agricultural crop (326 mm), primarily attributed to a reduction in ET. While the high flow rates (> 5 mm day^{-1}) increased from 2.3% to 2.6% (downstream) and 2.6% to 4.2% (upstream) for 25% to 50% conversion, the frequency was higher for the upstream location compared to the downstream location. These results were attributed to a substantial decrease in soil hydraulic conductivity in one of the dominant soils in the upstream location that is expected after land use conversion to agriculture. As a result, predicted subsurface drainage decreased, and surface runoff increased as soil hydraulic conductivity decreased for the soil upstream. The results indicate that soil hydraulic properties resulting from land use conversion have a greater influence on hydrologic components than the location of land use conversion. Wilk et al. (2001) calibrated a rainfall-runoff model for the period prior to conversion of the 12,200 km^2 Nam Pong watershed in Thailand where the forest area was reduced from 80% to 27% of the watershed. The calibrated model was used to predict outflows after the land use conversion. Siriwardena et al. (2006) used both forward and reverse modeling strategies to analyze impacts of clearing the natural forest cover on a 16,400 km^2 watershed in Queensland, Australia, over a relatively short period of time in the mid-1960s. Application of the calibrated models led to the conclusion that the clearing resulted in increased outflow by approximately 40%.

Chappell and Tych (2012) use dynamic harmonic regression (DHR) modeling of time series (i.e., daily streamflow) to separate step changes in forested watershed hydrology due to LUC from changes due to climatic cycles and shifts. The DHR defines a low frequency component to model trend, a periodic component to model seasonal variability, and a zero mean observation error component with a constant variance. The authors note that the disadvantage of the approach is such that hydrologic shifts due to changes in LULC may be masked by errors in observed data, seasonal and inter-annual cycles in the climatic data, and a slower rate of LUC (i.e., 20% clear-cutting versus 100% clear-cutting).

Simin et al. (2012) applied a Xinanjiang model-based change detection approach on a large 1,640 km^2 watershed in China. The authors reported that the runoff has declined by nearly 25% from 1976 to 2005 attributable to a decrease in medium- to high-coverage natural forest for expansion of tea gardens and human development.

Zégre et al. (2010) use the HBV-EC hydrologic model by Hamilton et al. (2000), generalized likelihood uncertainty estimation (GLUE), and generalized least squares (GLS) regression analysis to isolate effects of forest harvesting from variations in rainfall and streamflow as an alternative approach to the paired watershed approach. The latter approach is susceptible to erroneous change detection due to variability between the paired watersheds. The HBV-EC model is used as a virtual control in place of the control watershed. The model is calibrated using pre-treatment data and, subsequently, the calibrated model is used to simulate hydrologic variables during the post-treatment period. GLUE was used to address the uncertainty of simulated streamflow due to accrued errors in model structure, model identification, and input data. The GLS was used to detect change because it accounts for auto-correlation of the daily time series.

Using comprehensive global sensitivity analysis for DRAINMOD-FOREST, an integrated model for simulating water, carbon (C), nitrogen (N) cycling, and plant growth with the 21-year of data, Tian et al. (2013) demonstrated a need for of incorporating a dynamic plant growth model for simulating hydrological and biogeochemical processes in forest ecosystems, that would have ultimate implications on applying the model in assessing impacts of conversions of forests into other land uses.

In their long-term (1973–2000) SWAT simulation on a large degrading watershed in Kenya, Odira et al. (2010) found that with the expansion of the area under agriculture, the streamflow increased during the rainy seasons and reduced during the dry seasons, whereas when the area under forest cover was increased the peak streamflow reduced. When the forest cover is reduced to almost zero there was an increased peak and mean streamflow in the basin.

Using a validated water balance model, Kuchment et al. (2011) reported 30% larger snowmelt rates and 10 days on average longer duration of snowmelt after forest cutting in the northwestern part of Russia. Although the spring flood peaks were 50% lower and started 5–7 days later in the forested basin than the clearcut one, the simulated annual runoff appeared to be about 10% higher than the one with forest cutting as a result of snowmelt effects.

Continental-scale modeling allows for the examination of the spatial and temporal variability of hydrological response to LUC due to urbanization. Sun et al. (2011b) developed a monthly scale water balance model, called Water Supply Stress Index (WaSSI) model, by incorporating remote sensing data and a set of empirical ET models derived from global eddy flux measurements. Based on the WaSSI modeling, Caldwell et al. (2012) found that impervious cover increases total water yield when compared to native vegetation, and that the increase was most significant during the growing season in general (Fig. 7.2).

The proportion of streamflow that occurred as baseflow decreased somewhat, even though total water yield increased as a result of impervious cover. Water yield was most sensitive to changes in impervious cover in areas where annual ET is high relative to precipitation (i.e., the southwestern states: Texas and Florida). Water yield was less sensitive in areas with low ET relative to precipitation (i.e., Pacific Northwest and Northeastern States). Additionally, water yield was most impacted when high ET land cover types (i.e., forests) were converted to impervious cover than when lower ET land cover types (i.e., grassland) were converted. Using projections of future impervious cover provided by the U.S. EPA Integrated Climate and Land Use Scenarios project, this study predicts that water yield in urban areas of the Southwest, Texas, and Florida will be the most impacted by 2050, in part because these areas are projected to have significant increases in impervious cover and their unique climate. This study suggests that maintaining vegetation ET in urbanizing watershed is important for reducing hydrologic impacts. At a regional scale, watershed management should consider the

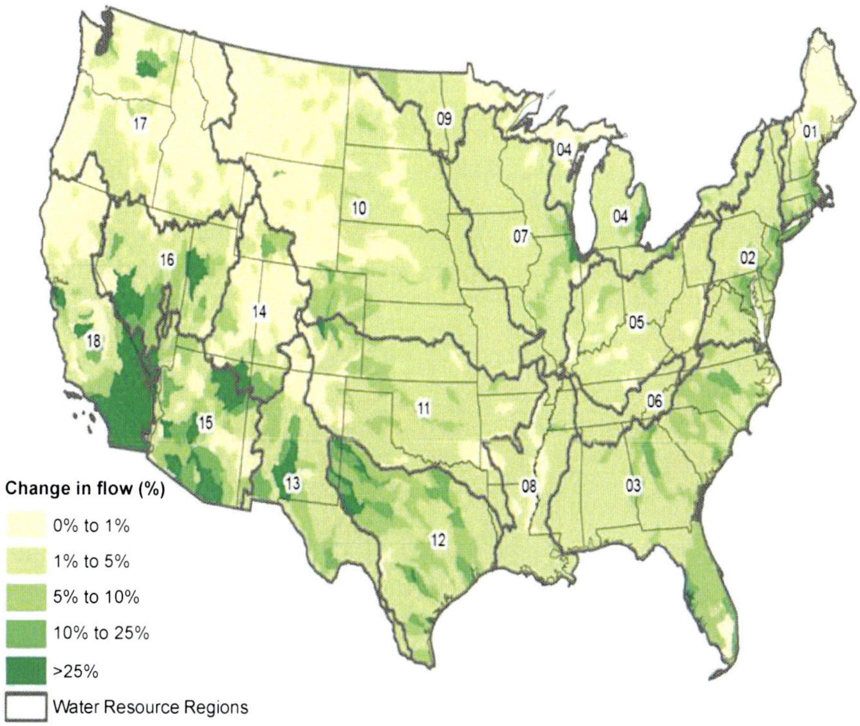

Figure 7.2. Simulated effects of impervious cover on annual water yield (%) in 2010 in the U.S. by a monthly scale water balance model, WaSSI (Caldwell et al., 2012).

climate-driven sensitivity of water yield to increases in impervious cover and the type of land cover being converted in addition to the magnitude of projected increases in impervious cover (Caldwell et al., 2012).

A SWAT simulation study with limited data in Arroio Lino in the state of Rio Grande do Sul in Brazil, that is covered by annual crops (tobacco) and forest at rates of approximately 30% and 50%, respectively, demonstrated that changed land uses impacted the hydrologic distribution (Bonumá et al., 2013). The SWAT model helped to demonstrate that baseflow is an important component of the area's hydrology and continued cropping on steep slopes would lead to greater erosion but also that a smaller scale model would possibly have caught the undulations in topography that the SWAT model missed. This led to an overestimation of hillslope runoff and; therefore, the water balance was incorrect.

Beyond multiple linear regressions for developing empirical relationships

Poor and Ullman (2010) indicate that statistical models developed using regression trees better predicted stream concentrations of nitrate and chloride than use of multiple linear regression in the Willamette River Basin (Oregon, U.S.). The predictive power as quantified by the coefficient of determination (R^2) increased from 0.35 to 0.75 for nitrate simulations and from 0.6 to 0.9 for chloride simulations. Their findings were consistent with earlier work by Creed et al. (2002) whose regression tree model explained 67% of the variability of observed soil nitrogen compared to 23% by the multiple linear regression model at Turkey Lakes Watershed, in central Ontario, Canada. However, Agren et al. (2010) show that if the independent variables are not highly correlated, multiple linear regression is still a powerful predictive tool for developing empirical relationships that simulate transport of nutrients in forested watersheds. They used readily accessible baseflow chemistry data (dissolved organic carbon [DOC]) and watershed landscape characteristics as explanatory variables to predict concentrations of DOC during peak flood conditions. Their parsimonious model comprised of log-transformed baseflow DOC, runoff, and percent wetlands; it explained 87% of total variability of the log-transformed peak flood DOC. They used principal component regression analysis to verify that covariance between explanatory variables did not influence the developed multiple linear regression model. In another study Amatya et al. (2007) compared four different empirical methods to estimate ET based on annual precipitation, PET, and watershed characteristics such as landcover, plant-water available coefficient, elevation, slope, and latitude that was used to estimate/validate annual streamflow for a forested watershed in coastal South Carolina.

Hydrologic recovery after disturbance

Sun et al. (2006) hypothesized that the impacts of forestation in China varied across five regions. The variation in rates of hydrologic recovery was attributed to effects of climate, soils, and vegetation. Troendle and King (1985) observed higher annual flows over a post-harvesting period of 30 years; however, the storm event peak flows over the same period were lower than pre-harvesting peak flows in some of the years. Their findings demonstrate how the effects of harvesting on different hydrologic responses (i.e., annual water yield, peak flow, and water table elevation) may be mediated by other factors such as rainfall intensity and duration. A review of hydrologic recovery on three watersheds in the Northeastern U.S. (Fernow, Leading Ridge, and Hubbard Brook) by Andreassian (2004) demonstrated that it took seven to 25 years for the annual water yields to return to pre-harvesting volumes. The difference in hydrologic recovery rate was attributed to difference in the percent composition of tree species that regrew after 100% clear-cutting (i.e., conversion of pre-treatment hardwood to post-treatment coniferous). Further review of the original work by Hornbeck et al. (1993) indicates that prolonged hydrologic recovery on the above three watersheds was due to application of herbicides to control natural regrowth. For watersheds where natural regrowth was not controlled (Hubbard Brook Catchments 4 and 5), pre-treatment annual water yields were attained within three to four years after harvesting (Hornbeck et al., 1986; Hornbeck et al., 1993; Hornbeck et al., 1997). Analysis by Swank et al. (2001) of 20 year annual water yields following clear-cutting of mixed hardwood forest in the southern Appalachian Mountains (Coweeta Hydrologic

Laboratory, USA), showed hydrologic recovery after five years. This observation was attributed to the rapid regrowth of the herbaceous species in the first three years after clear-cutting. Sun et al. (2000) showed spatial and temporal variation in the hydrologic response of forest clear-cutting on the southeastern coastal flatwoods of Florida. For example, clear-cutting of wetlands and uplands significantly increased the water table elevations. However, clearing cutting of only wetlands did not significantly affect water table elevations in the uplands. For both cases, the treatment effects were more pronounced during dry than wet periods. The same temporal (seasonal) effects were observed by Miwa (2004) and Blanton et al. (1998). A study by Grace et al. (2003) on hydrologic effects of harvesting a mainly hardwood forest stand in North Carolina (U.S.) indicated increases in event outflow, event peak flow and number of flow days. However, their results indicated no significant difference in daily outflow and daily water table depths about two years after harvest. The effect of forest harvesting on daily water table depth and outflow is contrary to results by Sun et al. (2001) for low gradient coastal watersheds. Grace et al. (2003) attributed the high variability of the two hydrologic responses to the record dry spell in 2001.

Scaling issues from plot to watershed scale and beyond

Spilsbury (2002) noted that the hydrological processes related with forests and water may be better understood at a plot level, or for a particular LULC type at a watershed scale but processes become more uncertain at greater spatial scales or smaller scales where the same processes operate across multiple land uses and management regimes. Liang et al. (2012) couple a landscape model (LANDIS) and an ecosystem process model (LINKAGES) to demonstrate that predictions of forest response to climate change at a watershed/landscape scale based on plot scale data are controlled by the sensitivity of forest species to heterogeneity of environmental factors such as temperature, precipitation, and soils. The study demonstrates that for forest species in a natural temperate forest (in China) that are highly sensitive to heterogeneity of environmental controls (i.e., Spruce and Birch), more experimental plots are required to accurately scale species distribution from plot to the watershed/landscape scale. While for species that are less sensitive to environmental controls (i.e., Larch and Fir), the choice of plot location is more instrumental than the number of experimental plots.

Bloschl (2001) suggested maybe, instead of trying to capture everything when upscaling, methods should be developed to identify dominant processes that control hydrological response in different environments and scales; then develop models to focus on these dominant processes, a notion that is called as the 'Dominant Processes Concept (DPC).' This may help with the generalization problems that have haunted hydrologists since the science began. Because most of the LUC effects are generally of interest and are assessed in the scales exceeding plot (< 0.1 ha) and field (< 100 ha) scales, some of the fine processes with less significant effects compared to others may be ignored for large scale assessments. That is how the models like SWAT (Arnold et al., 1998) has been developed and applied in large landscape scale assessments (Gassman et al., 2007; Arnold et al., 1998).

Remote sensing-based approach

Analysis of large complex landscapes with multiple land uses including forests and their effective management for sustainable development involves dealing with large scale spatial and tabular (attribute) data management. These data are becoming increasingly available worldwide with the advancement in satellite-based remote sensing technology. The need for assessing forest health by color changes due to chlorophyll loss has been stated by the United Nations REDD+ framework (Reducing Emissions from Deforestation and Forest Degradation). Carbon budget components such as biomass, dead organic matter, and soils need to be accounted for as well (Lynch et al., 2013). However, with the simultaneous advancements in high speed computing technology and geospatial technology, including geographical information systems (GIS), assessments of effects of land use conversion on various environmental and ecosystem functions are becoming possible for management decisions at various spatial levels from regional, continental, and global scales. Remote sensing uses various types of sensors such as multispectral, hyperspectral, ultraviolet, thermal sensors, light detection and ranging (LiDAR), radio detection and ranging (RADAR), and synthetic aperture radar (SAR), as well as others for collecting intricate information or attributes of forestry management (Franklin, 2001) including harvesting, plantation, and effects of disturbances like invasive species, hurricanes, and droughts. Similarly, time series or temporal analyses of satellite imageries and remote sensing data have been widely used for mapping, monitoring, and post-disturbance including LUC assessments (Pereira et al., 1997; Chuvieco et al., 2005; Chuvieco et al., 2007; Mitri and Gitas, 2010).

Researchers have found the utility of multitemporal medium/coarse satellite imagery from sensors such as Landsat, MODIS, SPOT-VGT and NOAA-AVHRR to assess fire severity (Veraverbeke et al., 2011) and monitor vegetation phenology and regrowth in burned areas (Goetz et al., 2006; Casady et al., 2010). Vegetation indices (VIs) such as NDVI and Soil Adjusted Vegetation Index (SAVI) developed by Huete (1988) were used in these post-fire monitoring studies and they provided accurate analysis on forest cover changes and classification over time. Besides these indices, remote sensing data are being used to derive the climatic (i.e., surface temperature, albedo) and vegetation (i.e., moisture, LAI, height) parameters for water balance, hydrologic, ecosystem, and climate change impact assessments. Panda et al. (2004) studied forest degradation using remote sensing and GIS in Indian forest ecosystems. They suggested that deforestation can be interpreted in terms of the conversion of forestland to other uses such as shifting agriculture (cropping or grazing followed by extended periods of fallow), permanent agriculture (cropping or grazing with little or no fallow), or urban uses. A comprehensive hydrological assessment study using data from a pair of gravity-measuring NASA satellites found that large parts of the arid Middle East region lost freshwater reserves during the past decade (www.jpl.nasa.gov/news/news.php?release=2013-054).

According to de Beurs and Henebry (2004), when change detection techniques using satellite images are based on short time series information, there is a greater risk that seasonal variation can be interpreted as change. For example, if the two different time periods in two different years are used in the analysis, the yearly vegetation cover changes (i.e., crop harvesting) may be considered as vegetation cover degradation. Therefore, caution should be taken while using remote sensing method for forest management and land use assessment studies because of the specific thresholds or change trajectories used in change detection for different spectral and phenological characteristics of land cover types (Lu et al., 2004; Verbesselt et al., 2009). Lynch et al. (2013) state that optical measurements should be taken every one to two weeks to achieve sufficient annual coverage to identify potential forest damage and possible warning signs for future prevention in detection of LULC. As reported by Verbesselt et al. (2009), a newly introduced method Breaks For Additive Seasonal and Trend (BFAST) approach enables the iterative decomposition of time series into trend, seasonal, and noise components resulting in the detection of gradual and abrupt changes in ecosystems and providing accurate data. Together, radar and optical systems can be used to create an early warning system that allows for daily scanning of forests thereby forming a 5–20 meter resolution that monitors logging in real time (Lynch et al., 2013). These authors state an alternative approach to this high cost method. A much less expensive choice is for cheaply made low-resolution optical satellites to monitor forests at a more sparse time scale (i.e., MODIS, DMC, SPOT, Landsat).

Effects of Various Factors in Land Use Change Impact Assessments

Thus far we have described the effects of deforestation and/or clear-cutting forests for land use conversion to agricultural crops or vice versa (i.e., afforestation or reforestation by planting forest on water yield, streamflows, peak flow rates, low flows, and water table dynamics). Almost all of these studies conclude and/or implicate this effect to reduce in ET as a result of canopy removal for deforestation/clear-cutting (Bosch and Hewlett, 1982; Brown et al., 2005). Similarly, the hydrologic effect of replacing pasture or other short crops with trees is reasonably well understood on a mean annual basis (Lane et al., 2003). Higher water yield from croplands/grasslands has been attributed to the lower ET from short crop/grass as compared to taller vegetation, which means that afforestation of grasslands would likely result in reductions in water yield. While it is true that ET is the largest component of the forest water balance, it is also a major component of the hydrologic cycle with direct impacts on water quantity, water quality, and net ecosystem and agri-ecosystem primary productivity. ET is influenced by parameters that vary across multiple scales—from site-specific variables such as soil, vegetation type, and localized weather conditions (PET), across the spatial heterogeneity of land use management at the landscape scales, to regional scales controlled by broad climatic conditions. Furthermore, the effects of LUC also depend on many other characteristics like water use and/or uptake (i.e., transpirational rates) by various vegetation types, percent imperviousness, soil types and hydraulic properties, and land management which may be even more complicated by climate change. There are several studies in the recent literature that estimate mean annual or annual ET as a difference of only annual precipitation and streamflow assuming no change in soil water storage (Lu et al., 2003; Sun et al., 2005; Amatya and Trettin, 2007; Amatya and Skaggs, 2011). Such estimates and other empirical models that include annual potential ET (PET) and vegetation factors (Sun et al., 2005; 2006; 2011b; Lu et al., 2003; Zhang et al., 2001; Turner et al., 1991; Calder and Newson, 1979) to assess streamflow for watersheds and its associated impacts. We synthesize below information on studies conducted to evaluate the effects of ET, methods of estimating ET, soil types and properties, land management, and climate change.

Effects of methods of estimating evapotranspiration

Most of the studies related to assessing effects of forest conversion to agricultural cropland have been conducted using hydrologic models. Breuer et al. (2009) reported that the magnitude of simulated effects depends substantially on the structure/method used in simulating ET. For example, in 10 models the authors evaluated, six of them used the Penman-Monteith method, one used Jensen-Haise method, one used Penman-Monteith, temperature driven monthly factors, one used solar radiation based, and the last one used an empirical temperature and precipitation driven method (Huisman et al., 2009). The authors concluded that although there was a general agreement among the models about the direction of changes in the mean annual discharge and 90% discharge, there was a considerable range in magnitude of predictions. Differences in the magnitude of flow increase were attributed to the different mean annual actual ET simulated by these models for each land use type. Similar findings were reported by Kim et al. (2012) who found the simulated drainage outflow sensitive to the method of estimating PET used in the DRAINWAT model. Similarly, Rao et al. (2011) examined three PET models (FAO P-M grass reference, Hamon, and Priestly Taylor) for possible applications in two mature forests in western North Carolina. The authors concluded that the first two models might underestimate the actual forest ET and thus might underestimate hydrologic effects of forest conversions. The Priestly-Taylor equation gave reasonable annual PET values, but applying the model to estimating actual ET requires calibration; it is unknown how the model performs at finer temporal scales since actual ET data are rarely available for forests. However, using three available PET methods in the SWAT model, Wang et al. (2006) reported that the AET values estimated by the three methods shared a concurrent spatial pattern and temporal trend and were insignificantly different from each other ($\alpha = 0.05$). The results indicated that after calibration, using the three ET methods within SWAT produced very similar hydrologic (AET and discharge) predictions for the studied watershed.

Gordon et al. (2005) as cited in Scanlon et al. (2007) reported a 4% reduction in global ET due to deforestation. Converting forested lands to the production of agricultural crops nearly always reduces ET and increases runoff (Skaggs et al., 1991; Skaggs et al., 2011; Sun et al., 2005; Amatya et al., 2008; Amatya and Trettin, 2007; Scanlon et al., 2007). Besides decreased ET, effects vary widely in quantity and in the timing based on the type of land, crop, and water management including the types of site preparation and the timing of such management during the year (Skaggs et al., 2011; Rab, 2004; Grace et al., 2006). In their study of assessing the impacts of reduction in forest cover on mean annual runoff using two empirical methods involving annual precipitation, land cover, elevation, and precipitation that fit the best with annual streamflow, Amatya and Trettin (2007) found an increase of as much as 62% runoff as a result of removal of 90% forest cover on the study watershed. Data in Table 7.1 present results from various types of studies that assess the effects of forest clearing and/or land use conversion from forests to agricultural croplands around the world.

Effects due to changes in type of crop, vegetation, and their water use

Forests generally have higher ET than other types of vegetation (1.6 times higher than grasslands (Zhang et al., 2001), as cited in Scanlon et al. (2007). In recent years, a need to better understand the relationship between watershed vegetation type and the variability of annual runoff as affected by vegetation manipulation for ET has found important implications for water resources management and development, stream ecology and fluvial geomorphology (Williams et al., 2012; Sun et al., 2005; De Wit, 2001). Holmes and Sinclair (1986) and Zhang et al. (2001) developed a relationship between annual ET and annual rainfall for various types of vegetation including grass and trees. Accordingly, in their study using worldwide fluxnet data, Williams et al. (2012) reported that grasslands on average have a higher evaporative index (ET/P) than forested landscapes, with 9% more annual precipitation consumed by annual ET compared to forests. The authors stated that while the Budyko framework's assumption of using mean annual precipitation and net radiation as two variables controlling mean annual ET and streamflow, vegetation type may well be another control. Brauman et al. (2012) also found that modeled PET from pasture was higher than that for the forest. This finding, according to the authors, was due to a balance between aerodynamically and stomatally controlled ET that differs significantly between two vegetation types, changing weighted sum of the two components yields, and lower PET at the forest sites.

Based on the SWAT model simulations, Schilling et al. (2008) concluded that historical LULC change in the U.S. Corn Belt region impacted the annual water balance in many Midwestern basins by decreasing annual ET and increasing streamflow and baseflow. Consistent with historical observations, their modeling results indicated increased corn production would decrease annual ET and increase water yield and losses of nitrate, phosphorus, and sediment, whereas increasing perennialization with grasses for ethanol biofuel would increase ET and decrease water yield and loss of nonpoint source pollutants. Global eddy flux ET data for different ecosystems have gradually become available for a general understanding of the environmental control of ET processes and validating hydrological models. Brauman et al. (2012) noted that concerns about reductions in water yield due to afforestation are likely to be relevant only in systems in which wind speeds are high and water stress limits ET.

Sun et al. (2011a) conducted a synthesis of ET studies for 13 worldwide intensively measured sites and found that monthly leaf area index (LAI) was the single most useful variable to explain ET variability across ecosystems over time, and PET and precipitation were additional key climatic variables for predicting monthly ET. There is a large variability in ET in space and time, and vegetation's influences on ET can be masked by climatic factors that are rather complex. Using similar eddy flux data from Ameriflux, a recent analysis suggests that forests use more water than grasslands

Table 7.1. Summary of studies on effects of land use change on streamflow.

Study	Site Name	Site Area, km²/Vegetation type	Method Used	Data/Simulation period	Mean annual rainfall/Runoff, mm	Increase or Decrease n Streamflow, mm (%)
Bonumà et al. (2013)	Rio Grande do Sul, Brazil	4.8 km^2/ Tobacco	SWAT model	2001–2005		Overpredicted sediments
Kim et al. (2012)	S4 coastal forest watershed, NC, U.S.	29.5/100% forest	DRAINWAT model	1951–2000	1290	65 (25%) increase for 100% conversion to cropland
Caldwell et al. (2012)	Continental scale for U.S.	Impervious area conditions in 2010 compared to past 30 years	WaSSI model	2010	Variable	mean 9.9% increase of flow in 2099 watersheds, median 2.2%
Kuchment et al. (2011)	North-western part of Russia		Water balance model			30% larger snowmelt rates; however, 10% higher runoff from forested than clearcut due to snowmelt
Odira et al. (2010)	Degrading watershed, Kenya		SWAT model	1973–2000		Streamflow (both peak and mean) increase in rainy season and decrease in dry season with increase in ag-land; increase in forest reduces peak flow
Qi et al. (2009)	Trent River watershed, Coastal NC	377/66% forest	USGS PRMS model	20 yrs (1981–01)	1300/426	59 (14% increase in flow)
Dai et al. (2009)	Control watershed, WS80, Coastal SC	1.6/100% forest	DHI- MIKESHE model	3 yrs (2003–06)	1270/269	113 (30% increase in flow)
Chescheir et al. (2009)	El Cerro basins, Uruguay	0.76 to 1.08	Paired watersheds	2000–2007		15% reduction for three years after planting
Dai et al. (2008)	Control watershed, WS80, Coastal SC	1.6/100% forest	DRAINMOD model	3 yrs (2003–06)	1270/269	122 (35% increase in flow)

Silverra and Alonso (2008)	Manual Diaz basin, Uruguay; Before and after afforestation (25% area)	2100	Measured data			8.2–36.5% flow decrease
Amatya and Trettin (2007)	Turkey Creek watershed, Coastal SC	72/96	EMPIRICAL: Rain, Canopy, PET	13 yrs (1964–76)	1320/350	208 (60% increase in flow)
Fernandez et al. (2007)	S4 watershed, Parker Tract, Coastal NC	30/50	DRAINMOD-based		1354/437	208 (16% increase in flow)
von Stackelberg et al. (2007)	Pastured watershed planted with pine, Uruguay	0.7–1.08	SWAT model	2000–2004	1487	23% reduction in mean annual flow
Siriwardena et al. (2006)	Natural forest, Queensland, Australia	16,400	Modeling startegy	Mid-1960s		Increase in outflow by 40%
Best et al. (2003)	Australia (pasture to pine and native forest to pasture;	0.94–3.10	Flow Duration Curve method	1974–1994	876	50% reduction in highflows and 100% reduction in low flows and for grass to pine and similar increase for vice versa; 30% reduction in New Zealnd
	New Zealand (grassland to pine)			1980–1999	1290–1310	
Wilk et al. (2001)	Nam Pong forested watershed, Thailand	12,200/80	Rainfall-runoff model			

Table 7.1. contd....

Table 7.1. contd.

Study	Site Name	Site Area, km²/Vegetation type	Method Used	Data/Simulation period	Mean annual rainfall/Runoff, mm	Increase or Decrease n Streamflow, mm (%)
Scott et al., 1998	South Africa watersheds; Grasslands planted to pine, eucalyptus, and wattle	2612–378,243 (1.2% forest)	Robust empirical model	Various lengths	448/37	3.2% reduction in mean annual flow
Jewitt (2002)						7.8% reduction in low flows
Scott and Smith (1997)	5 South African watersheds planted to pine and	0.26 to 2.0	Empirical relationships		1135–1611	Variable reductions with higher on Eucalyptus than the pine
	Eucalyptus				217–742	Higher percent reductions in low flows than in high flows
Dagg and Blackie (1965), Blackie (1972), Edwards et al. (1976)	Kenya & Tanzania:	0.544–0.702	Paired watershed approach and water balance	1958–1973	1500–2880	Longterm streamflow reduction of 8.9%
	Kericho (Montane forest to 54% tea estate) Kimakia (Vegetables to mature pine: 35 feet)					Water use at Kericho (Kenya) decreased by 14.4% after clearing and planting and increase by 2.4% with tea plantation
	Mbeya range (Evergreen forest to cultivated crops					
Dung et al. (2012)	Japan (43.2% thinning of a cypress forest)	0.002–0.004	Paired watershed approach	2004–2009	1732	240.7 mm increase in mean annual flow (36.7%) of 2 years

in general. The ET/P is much higher in deciduous forests than in grasslands. Irrigated croplands can have similar ET to forests (Fig. 7.3).

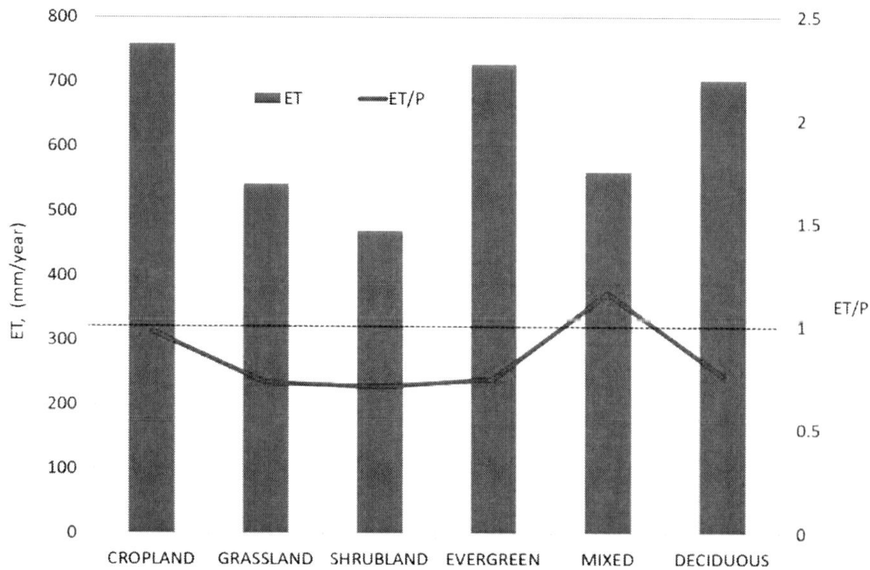

Figure 7.3. A comparison of annual ET and ET/P measured by the AmeriFlux network (Ge Sun, unpublished data).

Effects of land, soil and water management practices

In their simulation study on effects of land use on soil properties and hydrology of drained coastal plain watersheds using a validated DRAINMOD model, Skaggs et al. (2011) reported that the higher ET on an artificially drained pine forest resulted in reduced drainage outflow and deeper water table depth compared to an agricultural cropland site in North Carolina lower coastal plain in the U.S. The authors also argue that the assumption of approximation of soil-water properties based on soil type, independent of crop or land use, may not always be valid. For example, field effective hydraulic conductivity in the top 70 cm of the drained forest site was more than two orders of magnitude greater than that of the corresponding layers of soil on the agricultural site. Drainable porosity was much higher for the forested sites. As a result of these and large surface depressional storage, predicted surface runoff from the forested site was nil. Harvesting using heavy machines have the potential to disturb the forest soil surface including their structure and soil hydraulic properties (Rab, 2004; Skaggs et al., 2006; Grace et al., 2006). Skaggs et al. (2006) observed 20 to 30 times higher values of saturated conductivity (K_s) at 90 cm depth than those given in the NRCS Soil Survey Report for a Deloss fine sandy loam after harvest of a poorly drained mature pine plantation in coastal North Carolina. The authors noted that harvest did not appear to have affected those values, but site preparation for regeneration, including bedding, reduced the effective K values. A study by Grace et al. (2006) stated that soil compaction during harvesting of hardwoods on poorly drained soils of a Tidewater region of North Carolina increased the soil bulk density from 0.22 to 0.27 g cm^{-3} and decreased saturated hydraulic conductivity from 397 to 82 cm hr^{-1}. Rab (2004) reported that 10 years after timber harvesting in Australia, there was a 22–68% difference in soil physical properties of harvested compared to undisturbed soils.

Brauman et al. (2012) noted that in addition to afforestation, other changes in land use such as increased grazing intensity, can have an unexpected and pronounced role on soil properties. The authors reported while grazing reduces PET in temperate climates, an opposite effect was found in their humid site where short grazed grass increased PET rates. Grazing that reduces understory ET in forest may reduce water use. In a study on riparian deforestation effects, Greenberg et al. (2012) used LiDAR data to estimate changes in insolation that affect stream water temperatures and ecology.

Effects of climate change

The global climate change has direct (precipitation input, ET, and extreme events) and indirect (fires, insect disease, plant growth, invasive species) impacts on watershed hydrology, and has consequences of future LUC. In the southern U.S., climate warming is likely to increase ET, and thus decrease water yield if precipitation does not change (Sun et al., 2012). An increase in storm intensity and frequency is likely to increase in stormflow and peak flow rates and soil erosion potential (Sun et al., 2012; Dai et al., 2011). In such a case, the effects of deforestation or forest clearing for developments resulting in higher streamflows may be reduced due to higher soil and vegetation evaporation as a result of warming temperatures for prolonged growing season. At the same time seasonal flows may be further exacerbated due to projected increased intensity of storms. However, watershed water yield is most sensitive to precipitation change in a wet environment as shown in Fig. 7.4. Recently, Patterson et al. (2012) argued that whether the decrease in temperature with increase in observed precipitation and streamflow in the South Atlantic from 1964 to 1969 but with opposite trends from 1970 to 2005 have been driven by climatic or anthropogenic changes poses a great challenge to water resources managers.

Land use change and climate change do not occur independently. Lettenmaier et al. (1994) proposed that where streamflow does not follow climatic indicators, the cause is likely anthropogenic, although the LUC due to urbanization may confound climate-streamflow relationships (Shrestha et al., 2012). In an analysis of climate and streamflow data from six gauging stations from 1961 to 2006 in northeast China, Zhang et al. (2011) found that climate variability was estimated to account for 43% and human activities accounted for about 57%, respectively, of the reduction in the annual streamflow. Climate change often interacts with forest cover change to affect streamflow (Ford et al., 2011). Forest management may aggravate or mitigate climate change impacts on water yield. Urbanization generally increases stormflow and peakflow, impacts of which may likely be exacerbated due to extreme storm intensities projected as a result of climate change. Alternatively, climate change induced droughts

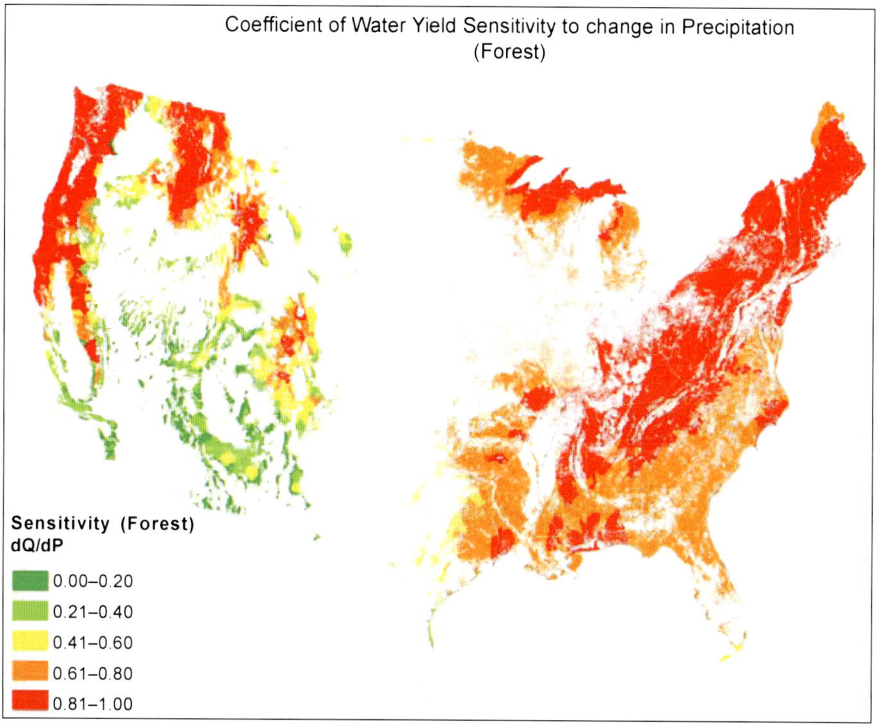

Figure 7.4. Forest water yield sensitivity (dQ/dP) to precipitation change as modeled using a method in Ma et al. (2008).

may aggravate the hydrologic influences of reforestation (i.e., water yield reduction). A parameter output sensitivity analysis suggests that with a reduction of leaf area index of 50% may double flow in the coastal plain in the humid southern U.S., but with minimal effects in the dry region (Fig. 7.5). In another study in an agricultural watershed on the Loess Plateau of China, Li et al. (2009) reported that the integrated effects of LUC and climate variability decreased runoff, soil water contents and ET. LUC increased ET by 8% while climate variability decreased by 103%. Similarly, a recent study by Shrestha et al. (2012) reported that the effects of changes in climatic variables on nutrient transport need to be considered with possible future changes in land use, crop type, fertilizer application, and transformation processes in the receiving water bodies.

Forests for bioenergy in a changing climate

The environmental value of forests will increase as climate change accelerates. While sustainably managed forests are encouraged for climate protection by carbon sequestration (UN General Assembly, 1994), the climate benefits of forest biogeophysical processes may equal those of carbon in the tropics (Bonan, 2008; Betts,

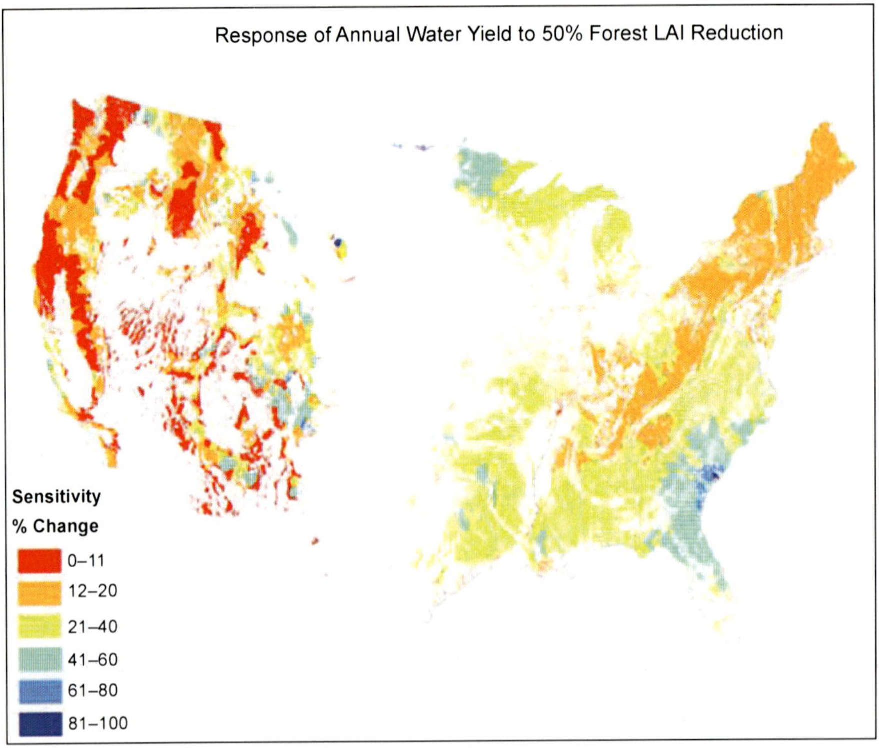

Figure 7.5. WaSSI modeled sensitivity of water yield to 50% reduction of forest leaf area index (LAI).

2011), although this diminishes in temperate and boreal forests. Cellulosic biofuel from forest-based bioenergy crops—whether short rotation woody crops, energy grass intercropping, or higher utilization of harvests—can increase again the value of forests and significantly reduce greenhouse gas emissions by replacing fossil fuels.

Human and natural disturbances are expected to become increasingly common; these alterations will impact forests (Dale et al., 2001), and be mitigated by them. By restricting pollutant movement due to vegetative ground cover, providing protection as a wind barrier, or acting as a filter by which chemicals are retained by the soil and/or organic debris or are remediated into alternative byproducts over time, forests are an *in situ* alternative to chemical treatment. Forests have been shown to intercept harmful radioactive elements (Kato et al., 2012), phytostabilize various metals and salts (shallow and deep rooted trees), and mineralize organic pollutants to carbon dioxide and salts. Trees reduce mass wasting and water held in fog moisture can be captured a increasing water resources in montane areas. Riparian forests protect watercourses from upland chemical uses (pesticides, fertilizers) and as a filter and depository for sediment (USFS, 2013). Forests cool nearby air and water bodies through shading and evaporation, reducing their temperature during hot season, providing a favorable habitat for aquatic life.

Sustainably managed forests operate under prescriptions that maximize economic return while protecting environmental resources. Many planning and operational guidelines are for water, with the majority being for water quality. BMPs have been shown to protect water resources, including aquatic life, by practices that vary in detail and implementation from state to state (Ice et al., 2010) but work as a system (Ice and Stednick, 2004). Minimizing net greenhouse gas emissions from forests and forest products may reprioritize forest benefits in a way that forces changes in BMPs. For example, current prescriptions call for limiting stream crossings, minimizing operational tract size, and disconnecting harvests. These inefficiencies in the harvesting process increase fuel consumption, and may need to be re-evaluated through the carbon life cycle analysis in order to provide the maximum benefit for biofuel solutions.

Several forest biogeophysical processes interact with climate, and are especially affected by albedo and vegetative ET. Strong forest ET cycles cool the atmosphere, while relatively low albedo (compared to agriculture) can have a warming effect. Arora and Montenegro (2011) introduce the concept of the "temperature effectiveness of afforestation" to aggregate the multiple effects of forests on climate, and demonstrate that forests can have either positive or negative effects on temperature, dependent on the characteristics of local climatic regime. Climate models predict that wetter areas will get wetter, although PET may outpace precipitation. In addition, storminess (Muschinski and Katz, 2013) and rainfall variability show signs of increase. More planted forests in wet areas could stabilize water yield, in addition to providing cooling, but forests in drier areas will be subject to die off due to great drought stress under climate change (Williams et al., 2013). These site-specific considerations will only become more important for intensive forest-based biofuel.

In addition to direct climate effects, the substitution of sustainably produced biofuel for fossil fuel can reduce greenhouse gas emission. Kior, the first commercial-scale producer of cellulosic biofuel, reports a reduction of 80% in emissions from their product made from wood chips over fossil-fuel emissions (Kior, 2013). While first-generation biofuels reduce emissions, the rain fed water use is a huge advantage over irrigated crops now often used for biofuel. The economic benefits of a shorter rotation crop may make conversion of marginal agricultural land feasible, moving land use from annual to perennial plantings and reducing erosion rates. Energy grasses have high water use efficiency and after establishment will protect soil from erosion. Higher harvest utilization takes advantage of an existing source of biomass. All of these possibilities are dependent on technology and economic conditions; however, research needs to continue into all possible scenarios to allow society to make the best choices that protect local environmental conditions and prioritize concerns.

A Conceptual Model

Based on our synthesis of information presented above on hydrologic impacts of forest removal as well as afforestation and effects of various factors including climate variability and change, a conceptual model was built with various types of stressors like LUC, climate change, and others interacting with various eco-hydrologic processes and functions with overall impacts on water quantity, streamflow timing and distribution, and water quality (Fig. 7.6).

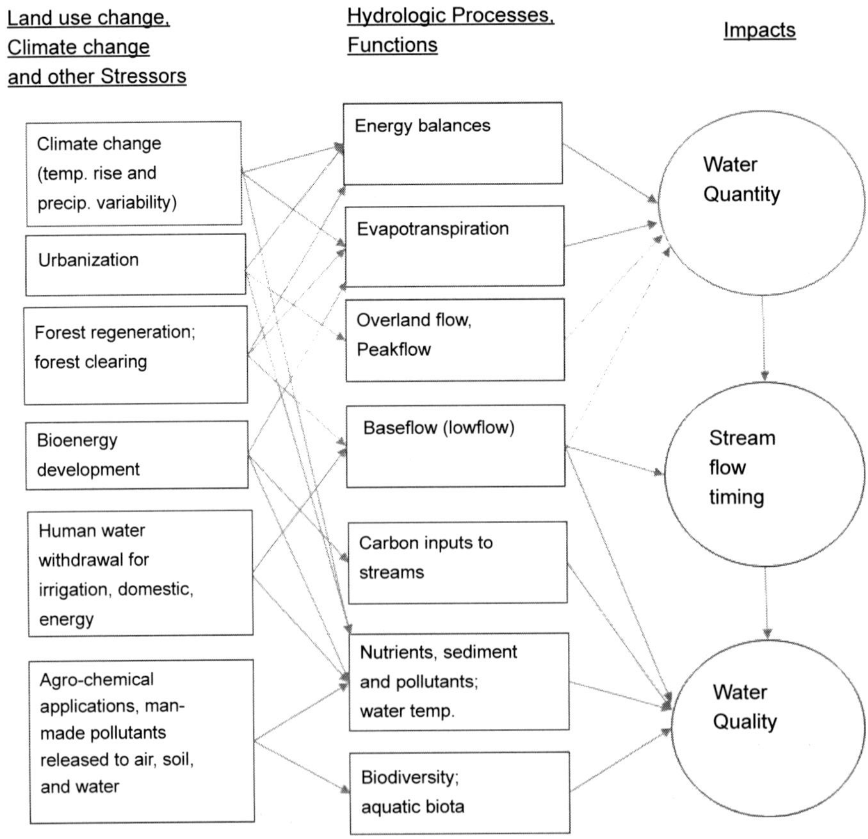

Figure 7.6. Factors affecting quantity, quality, and timing of streamflow.

Future Directions in Advancing Understanding the Interactions of Forests, Water, and Land Use Change

Water from forested landscapes is critical for supporting ecological systems and surrounding communities. Many factors that affect water quantity and quality, including LUC and climate change are summarized in a conceptual model (Fig. 7.6) that can help resource managers understand the interactions and pressures of various influences. Forest hydrology must advance to address current complex issues, including climate change, wildfires, changing patterns of development and ownership, and changing societal values (NRC, 2008; Jones et al., 2009). Sound science is needed to support wise decisions and appropriate responses to contentious water policy issues (USFS, 2011). The complex water and forest issues must be addressed using a 'one land' or a landscape approach. There is an urgent need for better understanding of the interface between forests/trees, and water, for awareness raising and capacity building in forest hydrology, and for embedding this knowledge and research findings in policies (Hamilton et al., 2008).

Future forest hydrologic science and related studies should move beyond 'forests'. Modern forest hydrology research should address issues on multiple spatial scales covering multiple landuses, including larger-scale contemporary water resource issues (i.e., water supply, instream flow, floods, droughts, beneficial water uses), climate change, hydrologic processes under disturbances (i.e., invasive species, fire, etc.) and urbanization (Sun et al., 2008; Amatya et al., 2011; Dai et al., 2011). Future forest land conversions and changing climate have a series of cascading effects (i.e., invasive species, fire, etc.) besides eco-hydrologic processes (Fig. 7.6; Vose et al., 2011; Brauman et al., 2012). To better understand their natural variability and for accurate impact assessments in field and landscape scales, longterm measurements at forest landscapes of varying ecological regions/characteristics accompanied by innovative monitoring technologies including remote sensing and advanced modeling are critical.

Defries and Eshelman (2004) call for a multidisciplinary approach with a comprehensive view towards the hydrologic processes that maintain ecological health and human requirements for food, water, and shelter. While the authors elaborated on available methods for assessing the effects of LUC from controlled experiments like paired watershed approach and mathematical modeling and their strengths and limitations, they expected that experimental approaches combining measurements from paired watersheds with process modeling will serve to unravel rapidly the response to LUC of watersheds of varying size, topography, and spatial configuration. Generally modelers calibrate and validate a model for current conditions of a watershed/landscape and predict the conditions of future treatment or calibrate the model for the first few years for pre-treatment and validate for the remaining time period for the treatment on the same watershed (Skaggs et al., 2011; Kuchment et al., 2011). A more novel approach for an accurate assessment of hydrologic and water quality impacts is to simulate, not only the reference conditions, but also the processes and interactions involved after the treatments. In that context, the applicability of the paired watershed approach has also been questioned in accurately predicting the peak flow rates, particularly their frequency and return periods (Kuras et al., 2012; Alila et al., 2009). Alila et al. (2009) argued that the science of forests and floods is in an urgent need of reevaluation of past studies in light of changing climates, insect epidemics, logging, and deforestation worldwide.

Spilsbury (2002) noted that most watershed management projects give little priority to research and monitoring, even though these are essential to effective watershed management and essential for establishing the efficacy of water management interventions. The author, therefore, concluded that future research efforts should focus on ways to maximize provision of environmental services in mixed land use mosaics, and strive to influence and inform public policy debates.

A scientific investigation of the causes and consequences of LULC requires an interdisciplinary approach integrating both natural and social scientific methods, which has emerged as the new discipline of land-change science (Ellis and Pontius, 2010). For more than a century agricultural and biological engineers have provided major advances in science, engineering, and technology to increase food and fiber production to meet the demands of a rapidly growing global population. Much of our agricultural land base originates from historically forested lands (Amatya et al., 2009), which have experienced dramatic declines and resurgence over the past century, including the Southern U.S. (USFS, 2011). The resulting landscape is a mosaic of agricultural, forest and urban lands that may not be sustainable with respect to the expected goods and ecosystem services including water quantity and quality (Amatya et al., 2011).

Water and nutrient balances quantified using long-term hydrology and water quality data from forested watersheds with minimal anthropogenic disturbances can serve as a reference for assessing the effects of LUCs into croplands and/or urban areas (Amatya and Skaggs, 2011). In this context, additional research is needed to advance our current understanding of forest hydrologic processes, especially the detail of stream networks, topographic depressions, floodplain and wetland functions, preferential flow characteristics into and within the forest soil profile, shallow and deep water table influences, flow generation in low-gradient watersheds, and the ET process for various forest types and species including the understory, which has received limited attention in the literature. An accurate understanding of these processes in a reference forest system is critical to the evaluation of impacts of all disturbances to the system. Hamilton et al. (2008) provide recommendations for a number of special forest situations important for water resources and their management. Those situations include management of cloud forests, swamp forests, riparian forest buffers, headwater forests for clean drinking water, and other forests to minimize salinization, erosion, etc.

Recent advancements in electronic sensor/digital monitoring, mapping, and remote sensing technology together with computing speed should also be used as opportunities to address these complex ecologic, hydrologic, and biogeochemical processes during land use conversion in a changing climate. Shuttleworth (2012) stated that future research is also required to fully validate recent interactive vegetation models, perhaps using remote sensing data.

Lockaby et al. (2011) reported that water stress will likely increase significantly by 2050 under four climate change scenarios largely because higher temperatures will result in more water loss by ET and because of decreased precipitation in some areas. Williams et al. (2012) concluded that climate type and vegetation should be considered in assessing ET, when streamflow is being regarded. However, the degree to which we can estimate changes in vegetation type with new ET requirements are speculative at best. The same authors also concluded that water stress due to the combined effects of population and LUC will increase by an average of 10% in the southern U.S. by 2050. Additional research is needed to identify innovative solutions and methodologies for mitigating potential impacts of climate and LUC for sustainable management of water resources on large prior converted agricultural landscapes that include forested watersheds. Furthermore, research needs more accurate information about the quantification of relationships between ecosystem attributes and forest management, including biomass production and harvest in a multi-dimensional context (Loehle et al., 2009).

According to the 4th Assessment Report of the Intergovernmental Panel on Climate Change (IPCC), the links between water and climate change are undeniable, with water predicted to be the primary medium through which early climate change impacts will be felt by people, ecosystems, and economies (WWC, 2009). Moreover, these climate change impacts will compound other existing pressures on water resources such as population growth, LUC, and changes in consumption patterns. As a result, further research is warranted to determine whether the impacts on streamflow and/or water availability trends have been driven by climatic or anthropogenic effects (i.e., LUC) posing a greater challenge to water resource managers and planners.

Acknowledgements

This work is a collaborative effort among researchers from USDA Forest Service, USDOI Bureau of Land Management, University of Georgia, University of North Georgia, and Weyerhaeuser Company. The authors would like to acknowledge the editor of the book for inviting us to contribute this book chapter.

References

Abdelnour, A., Stieglitz, M., Pan, F. and McKane, R. 2011. Catchment hydrological responses to forest harvest amount and spatial pattern. Wat. Res. Res. 47: 18.

Ågren, A., Buffam, I., Bishop, K. and Laudon, H. 2010. Modeling stream dissolved organic carbon concentrations during spring flood in the boreal forest: A simple empirical approach for regional predictions. Journal of Geophysical Research. Vol. 115(G1).

Alila, Y., Kuras, P.K., Schnorbus, M. and Hudson, R. 2009. Forests and floods: a new paradigm sheds light on age-old controversies. Water Resources Res. 45(8): 1–24.

Amatya, D.M., Douglas-Mankin, K.R., Williams, T.M., Skaggs, R.W. and Nettles, J.E. 2011. Advances in forest hydrology: challenges and opportunities. Trans. ASABE 54(6): 2049–2056.

Amatya, D.M., Hyunwoo, K., Chescheir, G.M., Skaggs, R.W. and Nettles, J.E. 2008. Hydrologic effects of size and location of land use conversion from pine forest to agricultural crop on organic soil. *In*: Proceedings, 2008 International Peat Congress, Tullamore, Ireland, June 8–12, 2008.

Amatya, D.M. and Skaggs, R.W. 2011. Long-term hydrology and water quality of a drained pine plantation in North Carolina, USA. Trans. ASABE 54(6): 2087–2098.

Amatya, D.M., Skaggs, R.W. and Gregory, J.D. 1996. Effects of controlled drainage on the hydrology of drained pine plantations in the North Carolina coastal plain. J. Hydro. 181(1): 211–232.

Amatya, D.M., Skaggs, R.W., Blanton, C.D. and Gilliam, J.W. 2006. Hydrologic and water quality effects of harvesting and regeneration of a drained pine forest. In Proc. of the ASABE-Weyerhaeuser sponsored Int'l Conference on Hydrology and Management of Forested Wetlands, eds. Williams and Nettles, New Bern, NC, April 8–12.

Amatya, D.M., Skaggs, R.W. and Trettin, C.C. 2009. Advancing the science of forest hydrology: a challenge to agricultural and biological engineers. Resource 16(5): 10–11. St. Joseph, Mich.: ASABE.

Amatya, D.M. and Trettin, C. 2007. Annual ET of a Forested Wetland Watershed, SC. Paper # 07-2222, St. Joseph, MI: ASAB.

Andreassian, V. 2004. Water and forests: from historical controversy to scientific debate. J. Hydrol. 291: 1–27.

Andreassian, V., Lerat, J., Moine, N.L. and Perrin, C. 2012. Neighbors: Nature's own hydrologic models. J. Hydrol. 414-415: 49–58.

Arnold, J.G., Srinivasan, R., Muttiah, R.S. and Williams, J.R. 1998. Large area hydrological modeling and assessment. Part I: Model development. J. Americ. Water Resour. Assoc. 34(1): 73–89.

Arora, V.K. and Montenegro, A. 2011. Small temperature benefits provided by realistic afforestation efforts. Nature Geoscience 4: 514–518.

Bates, C.G. 1921. First results in the streamflow experiment, Wagon Wheel Gap, Colorado. J. Forestry 19(4): 402–408.

Best, A. 2003. Acritical review of paired catchment studies with reference to seasonal flows and climatic variability. CSIRO Land and Water and Murray-Darling Basin Commission.

Betts, R.A. 2011. Climate science: Afforestation cools more or less. Nature Geoscience 4: 504–505.

Blackie, J. 1972. Hydrological effects of a change in land use from tropical forest to tea plantation in Kenya. *In*: Symposium on the Results of Research on Representative and Experimental Basins. Vol. 2. IASH-Unesco, p. 312.

Blanton, C.D., Skaggs, R.W., Amatya, D.M. and Chescheir, G.M. 1998. Soil hydraulic property variations during harvest and regeneration of drained, coastal pine plantations. In Proc. ASAE Annual Int. Meeting, Orlando, FL pp. 12–16.

Blöschl, G. 2001. Scaling in hydrology. Hydrological Processes 15(4): 709–711.

Bonan, G.B. 2008. Forests and climate change: forcings, feedbacks, and the climate benefits of forests. Science 320:1444–1449.

Bonumá, N.B., Rossi, C.G., Arnold, J.G., Reichert, J.M. and Paiva, E.M.C.D. 2013. Hydrology evaluation of the soil and water assessment tool for a small watershed in Southern Brazil. J. Appl. Eng. Agric. (in press).

Boggs, J., Sun, G., Jones, D. and McNulty, S.G. 2012. Effect of soils on water quantity and quality in Piedmont forested headwater watersheds of North Carolina. J. of the Amer. Water Resou. Assoc. (JAWRA) 1–19.

Bosch, J.M. and Hewlett, J.D. 1982. A review of catchment experiments to determine the effect of vegetation changes on water yield and ET. J. Hydrol. 55(1-4): 3–23.

Brauman, K.A., Freyberg, D. and Daily, G.C. 2012. Potential ET from forest and pasture in the tropics: A Case Study in Kona, Hawaii. J. Hydrol. 440-441: 52–61.

Bren, L. and Lane, P. 2014. Optimal development of calibration equations for paired catchment projects. J. Hydro. 10.1016/j.jhydrol.2014.07.059, 720–731.

Bren, L. and Mcguire, D. 2012. Paired catchment experiments and forestry politics in Australia. *In*: Revisiting Experimental Catchment Studies in Forest Hydrology. Proceedings of a Workshop held during the XXV IUGG General Assembly in Melbourne, June -July 2011. IAHS Publication 353: 106-116.

Breuer, L., Huisman, J.A., Willems, P., Bormann, H., Bronstert, A., Croke, B.F.W., Frede, H.-G., Graff, T., Hubrechts, L., Jakeman, A.J., Kite, G., Lanini, J., Leavesley, G., Lettenmaier, D.P., Lindstrom, G., Seibert, J., Sivapalan, M. and Viney, N.R. 2009. Assessing the impact of LUC on hydrology by ensemble modeling (LUCHEM). I: Model intercomparison with current land use. Adv. Wat. Res. 32: 129–146.

Brown, T.C., Hobbins, M.T. and Ramirez, J.A. 2008. Spatial distribution of water supply in the coterminous United States. J. Amer. Water Resou. Assoc. (44)6: 1474–1487.

Brown, A.E., Zhang, L., McMahon, T.A., Western, A.W. and Vertessey, R.A. 2005. A review of paired catchment studies for determining changes in water yield resulting from alterations in vegetation. J. Hydrol. 310(1-4): 28–61.

Calder, I.R. and Newson, M.D. 1979. Land use and upland water resources in Britain—a strategic look. Water Resour. Bull. 16: 1628–1639.

Caldwell, P.V., Sun, G., McNulty, S.G., Cohen, E.C. and Moore Myers, J.A. 2012. Impacts of impervious cover, water withdrawals, and climate change on river flows in the Conterminous US, Hydrol. Earth Syst. Sci. 9: 4263–4304.

Casady, G., van Leeuwen, W. and Marsh, S. 2010. Evaluating post-wildfire vegetation regeneration as a response to multiple environmental determinants. Environ. Model Assess. 15(5): 295–307.

Chang, M. 2003. Forest Hydrology: An Introduction to Water and Forests. CRC Press LLC, Boca Raton, Florida.

Chappell, N.A. and Tych, W. 2012. Identifying step changes in single streamflow and evaporation records due to forest cover change. Hydrological Processes Vol. 26(1): 100–116.

Chappell, N.A. and Tych, W. 2012. Identifying step changes in single streamflow and evaporation records due to forest cover change. Hydrological Processes 26(1): 100–116.

Chescheir, G.M., Skaggs, R.W. and Amatya, D.M. 2009. Quantifying the hydrologic impacts of afforestation in Uruguay: a paired Watershed study. *In*: Proc., XIII World Forestry Congress, Buenos Aires, Argentina, 18–23 October.

Chuvieco, E., Englefield, P., Trishchenko, A.P. and Luo, Y. 2007. Generation of long time series of burn area maps of the boreal forest from NOAA-AVHRR composite data. Remote Sensing of Environment 112(5): 2381–2396. Earth Observations for Terrestrial Biodiversity and Ecosystems Special Issue, 15 May 2008.

Chuvieco, E., Ventura, G. and Martín, M.P. 2005. AVHRR multitemporal compositing techniques for burned land mapping. International Journal of Remote Sensing 26(5): 1013–1018.

Clifton, C., Daamen, C., Horne, A. and Sherwood, J. 2006. Water, LUC and 'new forests': what are the challenges for south-west Victoria? Australian Forestry 69(2): 95–100.

Creed, I.F., Trick, C.G., Band, L.E. and Morrison, I.K. 2002. Characterizing the spatial pattern of soil carbon and nitrogen pools in the Turkey Lakes Watershed: a comparison of regression techniques. Water, Air, and Soil Pollution 2: 81–102.

Dagg, M. and Blackie, J. 1965. Studies of the effects of changes in land use on the hydrological cycle in East Africa by means of experimental catchment areas. Hydrological Sciences J. 10(4): 63–75.

Dai, Z., Trettin, C.C. and Amatya, D.M. 2013. Effects of climate variability on forest hydrology and carbon sequestration on the Santee Experimental Forest in Coastal South Carolina. USDA Forest Service South. Res. Station, Gen. Tech. Rep. SRS-172, 32 p.

Dai, Z., Amatya, D.M., Sun, G., Li, C., Trettin, C.C. and Li, H. 2008. Modeling effects of LUC on hydrology of a forested watershed in coastal South Carolina. Proceedings of the 2008 South Carolina Water Resources Conference, held October 14-15, 2008 at the Charleston Area Event Center.

http://www.clemson.edu/restoration/events/past_events/sc_water_resources/t3_proceedings_presentations/t3_zip/daiz.pdf.

Dai, Z., Amatya, D., Sun, G., Trettin, C., Li, C. and Li, H. 2011. Climate variability and its impact on forest hydrology on South Carolina Coastal Plain of USA. Atmosphere 2: 330–357.

Dai, Z., Trettin, C., Li, C., Amatya, D., Sun, G. and Li, H. 2009. A comparison of two different scales of hydrologic models for modeling forested wetland watershed hydrology in Coastal South Carolina, USA. Paper # 09-6656, 2009 ASABE Annual International meeting, Reno, NV, June 21–24.

Dale, V.A., Joyce, L.A., McNulty, S., Neilson, R.P., Ayres, M.P., Flannigan, M.D., Hanson, P.J., Irland, L.C., Lugo, A.E., Peterson, C.J., Simberloff, D., Swanson, R.J., Stocks, B.J. and Wotton, M. 2001. Climate Change and Forest Disturbances. Bioscience 51(9): 723–734.

De Beurs, K.M. and Henebry, G.M. 2004. Land surface phenology, climatic variation, and institutional change: analyzing agricultural land cover change in Kazakhstan. Remote Sens. Environ. 89: 497–509.

DeFries, R. and Eshleman, K.N. 2004. Land-use change and hydrologic processes: a major focus for the future. Hydrol. Process 18: 2183–2186.

De la Cretaz, A. and Barten, P.K. 2007. Land Use Effects on Streamflow and Water Quality in Northeastern United States. CRC Press, Boca Raton, FL.

De Wit, M.J.M. 2001. Nutrient fluxes at the river basin scale. I: the PolFlow model. Hydro. Proc. 15: 743–759.

Dung, B.X., Gomi, T., Miyata, S., Sidle, R.C., Kosugi, K. and Onda, Y. 2012. Runoff responses to forest thinning at plot and catchment scales in a headwater catchment draining Japanese cypress forest. J. Hydro. 444: 51–62.

Edwards, K. and Blackie, J. 1981. Results of the East African catchment experiments 1958–1974. Tropical Agricultural Hydrology 163–188.

Edwards, K., Blackie, J. and Eeles, C. 1976. Final Report of the East African Catchment Research Project (ODM R2582). Vol. 2: The Kericho Experiments.

Edwards, P. and Troendle, C.A. 2008. Water yield and hydrology. *In*: Audin, L.J. (ed.). Cumulative Watershed Effects of Fuels Management. Newton Square, Pa.: U.S. Forest Service, Northeastern Area. Available at: www.na.fs.fed.us/fire/cwe.shtm. Accessed 15 June 2009.

Elliott, K.J. and Vose, J.M. 2005. Initial effects of prescribed fire on quality of soil solution and streamwater in the southern Appalachian Mountains. Southern J. Appl. Forestry 29(1): 5–15.

Ellis, E. and Pontius, R. 2010. Land-use and land-cover change. *In*: Encyclopedia of Earth. Eds. Cutler J. Cleveland (Washington, D.C.: Environmental Information Coalition, National Council for Science and the Environment) [First published in the Encyclopedia of Earth April 18, 2010; Last revised Date May 7, 2012; Retrieved March 1, 2013 <http://www.eoearth.org/article/Land-use_and_land-cover_change.

EOS. 2012. The Pros and Cos of Trading Water: A Case Study in Australia. In AGU's EOS, 93(47), p 496; Excerpt from Water Resour. Research.

Fahey, B. and Jackson, R. 1997. Hydrological Impacts of converting native forests and grasslands to pine plantations, South Island, New Zealand. Agric. and For. Meteorology 84(1997): 69–82.

FAO. 2013. Forests and Water: International Momentum and Action. Forestry Department, Food and Agriculture Organization of the United Nations (FAO), Rome, Italy, 2013.

Farley, K.A., Jobbagy, I.E. and Jackson, R.B. 2005. Effects of afforestation on water yield: a global synthesis with implications for policy. Glob. Change Biol. 11: 1565–1576.

Fernandez, G.P., Chescheir, G.M., Skaggs, R.W. and Amatya, D.M. 2007. Application of DRAINMOD-GIS to a lower coastal plain watershed. Trans. Am. Soc. Agric. Bio. Engrs. 50(2): 439–447.

Ford, C.R., Laseter, S.H., Swank, W.T. and Vose, J.M. 2011. Can forest management be used to sustain water-based ecosystem services in the face of climate change? Ecological Applications 21(6): 2049–2067.

"Forest Types of the World-Maps of the World's Forest". Forestry.about.com. Last accessed 27 Feb, 2013.

Franklin, S.E. 2001. Remote sensing for sustainable forest management. Lewis publishers, DC.

Gassman, P.W., Reyes, M.R., Green, C.H. and Arnold, J.G. 2007. The soil and water assessment tool: historical development, applications, and future research directions. Trans. ASABE 50(4): 1211–1250.

GEF. 2012. Land use, land-use change, and forestry (LULUCF) activities. (Authored by Burke, Marianne, Karan Chouksey, and Linda Heath.) Global Environment Facility, Washington, DC. Available at http://

www.thegef.org/gef/pubs/land-use-land-use-change-and-forestry-lulucf-activities (Last accessed 21 Feb 2013). Profes. Graph. Printing Co. ISBN: 978-1-939339-47-8.

Goetz, S., Fiske, G. and Bunn, A. 2006. Using satellite time-series data sets to analyze fire disturbance and forest recovery across Canada. Remote Sensing of Environment 92: 411–423.

Gordon, L.J., Steffen, W., Jonsson, B.F., Folke, C., Falkenmark, M. and Johannessen, A. 2005. Human modification of global water vapor flows from the land surface. Proc. Natl. Acad. Sci. U.S.A. 102: 7612–7617.

Grace, J.M., Skaggs, R.W. and Cassel, D.K. 2006. Soil physical changes associated with forest harvesting operations on an organic soil. SSSAJ 70(2): 503–509.

Grace, J.M., III, Skaggs, R.W., Malcom, H.R., Chescheir, G.M. and Cassel, D.K. 2003. Increased water yields following harvesting operations on a drained coastal watershed. Proceedings of 2003 ASAE Annual International Meeting.

Green, C.H., Tomer, M.D., Di Luzio, M. and Arnold, J.G. 2006. Hydrologic evaluation of the Soil and Water Assessment Tool for a large tile-drained watershed in Iowa. Trans. ASABE 49: 413–422.

Greenberg, J.A., Hestir, E.L., Riano, D., Scheer, G.J. and Ustin, S.L. 2012. Using LiDAR data analysis to estimate changes in insolation under large-scale riparian deforestation. J. Amer. Wat. Resou. Assoc. 48(5): 939–948.

Hamilton, A.S., Hutchinson, D.G. and Moore, R.D. 2000. Estimating winter streamflow using conceptual streamflow model. J. Cold. Reg. Eng. 14(4): 158–175.

Hamilton, L.S., Dudley, N., Greminger, G., Hassan, N., Lamb, D., Stolton, S. and Tognetti, S. 2008. Forests and Water. FAO Forestry Paper 155, Food and Agriculture Organization of the United Nations, Rome.

Heuvelmans, G., Muys, B. and Feyen, J. 2004. Evaluation of hydrological model parameter transferability for simulating the impact of land use on catchment hydrology. Physics and Chem. of the Earth A/B/C 29(11-12): 739–747.

Holmes, J.W. and Sinclair, J.A. 1986. Water yield from some afforested catchments in Victoria. Hydrology and Water Resources Symposium, Griffith University, Brisbane, National Conference Publication 86/13. Canberra, Australia: Institution of Engineers.

Hornbeck, J., Adams, M., Corbett, E., Verry, E. and Lynch, J. 1993. Long-term impacts of forest treatments on water yield: a summary for northeastern U.S.A., J. Hydrol. 150(2-4): 323–344.

Hornbeck, J., Martin, C. and Eagar, C. 1997. Summary of water yield experiments at Hubbard Brook experimental forest, New Hampshire, Canadian J. Forest Research 27(12): 2043–2052.

Hornbeck, J., Martin, C., Pierce, R., Bormann, F., Likens, G. and Eaton, J. 1986. Clearcutting northern hardwoods: effects on hydrologic and nutrient ion budgets, Forest Science 32(3): 667–686.

Huete, A.R. 1988. A soil-adjusted vegetation index (SAVI). Remote Sensing of Environment. 25: 295–309.

Huisman, J.A., Bruer, L., Bormann, H., Bronstert, A., Croke, B.F.W., Frede, H.-G., Graff, T., Hubrechts, L., Jakeman, A.J., Kite, G., Lanini, J., Leavesley, G., Lettenmaier, D.P., Lindstrom, G., Seibert, J., Sivapalan, M., Viney, N.R. and Willems, P. 2009. Assessing the impact of LUC on hydrology by ensemble modeling (LUCHEM). II: Scenario Analysis. Adv. Wat. Res. 32: 159–170.

Ice, G.G., Schilling, E. and Vowell, J. 2010. Trends for forestry best management practices implementation. J. Forestry 108(6): 267–273.

Ice, G.G. and Stednick, J.D. (eds.). 2004. A Century of Forest and Wildland Watershed Lessons. Society of American Foresters, Bethesda, Md.

Jackson, C.R., Sun, G., Amatya, D.M., Swank, W.T., Riedel, M., Patric, J., Williams, T., Vose, J.M., Trettin, K., Aust, W.M., Beasely, S., Williston, H. and Ice, G. 2004. Fifty years of forest hydrology research in the southeast: some lessons learned. pp. 33–112. *In*: Stednick, J. and Ice, G. (eds.). A Century of Forest and Wildland Watershed Lessons. Society of American Foresters, Bethesda, Maryland, U.S.A.

Jayakaran, A., Williams, T.M., Ssegane, H.S., Amatya, D.M., Song, B. and Trettin, C. 2014. Hurricane impacts on a pair of coastal forested watersheds: implications of selective hurricane damage to forest structure and streamflow dynamics. Hydrol. Earth Syst. Sci. (HESS) 18: 1151–1164.

Jewitt, G. 2002. Guest Editorial The 8%-4% debate: Commercial afforestation and water use in South Africa. Southern Forests: a Journal of Forest Science 194(1): 1–5.

Jones, J.A., Achterman, G.L., Augustine, L.A., Creed, I.F., Ffolliott, P.F., MacDonald, L. and Wemple, B.C. 2009. Hydrologic effects of a changing forested landscape: challenges for hydrological sciences. Hydrol. Proc. 23(18): 2699–2704.

Kato, H., Onda, Y. and Gomi, T. 2012. Interception of the Fukushima reactor accident-derived 137Cs, 134Cs and 131I by coniferous forest canopies. Geophysical Research Letters, doi:10.1029/2012GL052928.

Kim, H.W., Amatya, D.M., Skaggs, R.W. and Chescheir, G.M. 2012. Hydrologic effects of size and location of fields converted from drained pine forest to agricultural cropland. J. Hydrologic Engineering. Accepted December 28, 2011; posted ahead of print January 2, 2012.

King, J.S., Ceulemans, R., Albaugh, J.M., Dillen, S.Y., Domec, J.-C., Fichot, R., Fischer, M., Leggett, Z., Sucre, E., Trnka, M. and Zenone, T. 2013. The challenge of lignocellulosic bioenergy in a water-limited world. BioScience 63: 102–117.

Kior. 2013. Citing report from TIAX, LLC. Retrieved from http://www.kior.com/content/?s=5&s2=23&p=23&t=Greenhouse-Gas-Reductions.

Kuchment, L.S., Gelfan, A.N. and Demidov, V.N. 2011. Modeling of the hydrologic cycle of a forest river basin and hydrological consequences of forest cutting. The Open Hydrol. J. 5: 9–18.

Kuras, P.K., Alila, Y. and Weiler, M. 2012. Forest harvesting effects on the magnitude and frequency of peak flows can increase with return periods. Wat. Resour. Res. 48: 1–19.

Lane, P., Best, A., Hickel, K. and Zhang, L. 2003. The effect of afforestation on flow duration curves. Technical report, report 03/13, November 2003, Cooperative Research Center for Catchment Hydrology, Monash University, VIC 300, Australia, 26 p.

Lettenmaier, D.P., Wood, E.F. and Wallis, J.R. 1994. Hydro-Climatological trends in the Continental United States, 1948–88. J. Climate 7: 586–607.

Laur´en, A., Heinonen, J., Koivusalo, H., Sarkkola, S., Tattari, S., Mattsson, T., Ahtiainen, M., Joen-suu, S., Kokkonen, T. and Fin´er, L. 2009. Implications of uncertainty in a pre-treatment dataset when estimating treatment effects in paired catchment studies: phosphorus loads from forest clear-cuts. Water, Air, & Soil Pollution 196(1): 251–261.

Li, Z., Liu, W.Z., Zhang, X.C. and Zheng, F.L. 2009. Impacts of LUC and climate variability on hydrology in an agricultural catchment on the Loess Plateau of China. J. Hydrol. 377: 3–42.

Liang, Y., He, S.H., Yang, J. and Wu, Z.W. 2012. Coupling ecosystem and landscape models to study the effects of plot number and location on prediction of forest landscape change. Landscape Ecology (2012)27: 1031–1044. DOI 10.1007/s10980-012-9759-7

Lockaby, G., Nagy, C., Vose, J.M., Ford, C.R., Sun, G., McNulty, S., Caldwell, P., Cohen, E. and Moore Myers, J.A. 2011. Water and Forests. *In*: Wear, D.N. and Greis, J.G. (eds.). The Southern Forest Futures Project: Technical Report, USDA Forest Service, Southern Research Station, Asheville, NC., General Technical Report, 2011.

Loehle, C., Wigley, T.B., Schilling, E., Tatum, V., Beebe, J., Vance, E., Deusen, P.V. and Weatherford, P. 2009. Achieving conservation goals in managed forests of the southeastern coastal plain. Environ. Mgmt. 44(6): 1136–1148.

Lorup, J.K., Refsgaard, J.C. and Mazvimavi, D. 1998. Assessing the effects of LUC on catchment runoff by combined use of statistical tests and hydrological modeling: Case studies from Zimbabwe. J. Hydrol. 205 (3-4): 147–163.

Lu, D., Mausel, P., Brondizio, E. and Moran, E. 2004. Change detection techniques. Int. J. Remote Sensing 25(12): 2365–2407.

Lu, J., Sun, G., McNulty, S.G. and Amatya, D.M. 2003. Modeling actual ET from forested watersheds across the southern United States. J. Amer. Wat. Res. Assoc. 39(4): 887–896.

Lynch, J., Maslin, M., Balzter, H. and Sweeting, M. 2013. Choose satellites to monitor deforestation. Nature 496: 293–294.

Ma, Z.M., Kang, S.Z., Zhang, L., Tong, L. and Su, X.L. 2008. Analysis of impacts of climate variability and human activity on streamflow for a river basin in arid region of northwest China. J. Hydrol. 352: 239–249.

Mitri, G. and Gitas, I.Z. 2010. Mapping postfire vegetation recovery using EO-1 Hyperion Imagery. IEEE Transactions on Geoscience and Remote Sensing 48(3): 1613–1618.

Miwa, M., Aust, W.M., Burger, J.A., Patterson, S.C. and Carter, E.A. 2004. Wet-weather timber harvesting and site preparation effects on coastal plain sites: a review, Notes.

Muschinski, T. and Katz, J.I. 2013. Trends in Hourly Rainfall Statistics in the United States under a Warming Climate. Nature Climate Change 1–4.

Neitsch, S.L., Arnold, J.G., Kiniry, J.R. and Williams, J.R. 2005. Soil and Water Assessment Tool Theoretical Documentation, Version 2005. Temple, TX.: USDA-ARS Grassland, Soil and Water Research Laboratory.

NRC. 2008. Hydrologic Effects of a Changing Forest Landscape. In "Free Executive Summary", National Research Council, National Academy of Sciences, The National Academies Press, Washington, D.C., ISBN: 978-0-309-12108-8, 180 p.

Odira, P.M.A., Nyadawa, M.O., Ndwallah, B.O., Juma, N.A. and Obiero, J.P. 2010. Impact of land use/cover dynamics on streamflow: a case of Nzoia River catchment, Kenya. Nile Basin Water Science & Engineering J. 3(2): 64–110.

Panda, S.S., Andrianasolo, H., Murty, V.V.N. and Nualchawee, K. 2004. Forest management planning for soil conservation using satellite images, GIS mapping, and soil erosion modeling. J. Environmental Hydrology 12(13): 1–16.

Patterson, L.A., Lutz, B. and Doyle, M.W. 2012. Streamflow changes in the South Atlantic, United States during the mid- and late 20th century. J. Amer. Wat. Resou. Assoc. 48(6): 1126–1138.

Pereira, M., Chuvieco, E., Beudoin, A. and Desbois, N. 1997. Remote sensing of burned areas: a review. pp. 127–184. *In*: Chuvieco, E. (ed.). A Review of Remote Sensing Methods for the Study of Large Fires. Universidad de Alcala, Alcala de Henares, Spain, pp. 127–183.

Poor, C.J. and Ullman, J.L. 2010. Using regression tree analysis to improve predictions of low-flow nitrate and chloride in Willamette River Basin watersheds. Environmental Management 46(5): 771–780.

Prestemon, J. and Abt, R. 2002. Timber products supply and demand. pp. 299–325. *In*: Wear, D. and Greis, J. (eds.). Southern Forest Resource Assessment. U.S. Forest Service, Southern Research Station, Asheville, N.C.

Price, K., Jackson, C.R., Parker, A.J., Reitan, T., Dowd, J. and Cyterski, M. 2011. Effects of watershed land use and geomorphology on stream flows during severe drought conditions in the southern Blue Ridge Mountains, Georgia, and North Carolina, Unites States. Water Res. Res. Vol. 47.

Qi, S., Sun, G., Wang, Y., McNulty, S.G. and Myers, J.A.M. 2009. Streamflow response to climate and landuse changes in a coastal watershed in North Carolina. Trans. ASABE 52(3): 739–749.

Rab, M.A. 2004. Recovery of soil physical properties from compaction and soil profile disturbance caused by logging of native forest in Victorian central Highlands, Australia. For. Ecol. Manage. 191: 329–340.

Rao, L., Sun, G., Ford, C.R. and Vose, J. 2011. Modeling potential ET of two forested watersheds in the Southern Appalachians, USA. Trans. ASABE 54(6): 2067–2078.

Sahin, V. and Hall, M.J. 1996. The effects of afforestation and deforestation on water yields. J. Hydro. 178(1-4): 293–309.

Scanlon, B.R., Jolly, I., Sophocleous, M. and Zhang, L. 2007. Global impacts of conversions from natural to agricultural ecosystems on water resources: Quantity versus quality. Water Res. Res. 43.

Schilling, K.E., Jha, M.K., Zhang, Y.-K., Gassman, P.W. and Wolter, C.F. 2008. Impact of land use and land cover change on the water balance of a large agricultural watershed: historical effects and future directions. Water Resour. Res. Vol 45, Issue 7, 10.1029/2007WR006644.

Scott, D. and Lesch, W. 1997. Streamflow responses to afforestation with *Eucalyptus grandis* and Pinus patula and to felling in the Mokobulaan experimental catchments, South Africa. J. Hydro. 199(3): 360–377.

Scott, D., Maitre, D. and Fairbanks, D. 1998. Forestry and streamflow reductions in South Africa: A reference system for assessing extent and distribution. Water S.A. 24(3): 187–200.

Scott, D.F. and Smith, R.E. 1997. Preliminary empirical models to predict reductions in total and low flows resulting from afforestation. Water S.A. 23(2): 135–140.

Sedell, J., Sharpe, M., Apple, D.D., Copenhagen, M. and Furniss, M. 2000. Water and the Forest Service. Washington, D.C.: U.S. Forest Service.

Shrestha, R.R., Dibike, Y.B. and Prowse, T.D. 2012. Modeling climate change impacts on hydrology and nutrient loading in the Upper Assiniboine catchment. J. Amer. Wat. Resou. Assoc. 48(1): 74–89.

Shuttleworth, W.J. 2012. Terrestrial Hydrometeorology. Wiley Blackwell & Sons, Ltd., West Sussex, U.K.

Silveira, L. and Alonso, J. 2008. Runoff modifications due to the conversion of natural grasslands to forests in a large basin in Uruguay. Hydrol. Process 23: 320–329.

Simin, Q., Weimin, B., Peng, S., Zhongbo, Y., Peng, L., Bo, Z. and Peng, J. 2011. Evaluation of runoff responses to LUCs and land cover changes in the Upper Huaihe River basin, China. J. Hydrol. Engrg. 17(7): 800–806.

Singh, V.P. and Frevert, D.K. 2006. Watershed Models. CRC Press, Taylor & Francis Group, Boca Raton, FL.

Siriwardena, L., Finlayson, B.L. and McMahon, T.A. 2006. The impact of LUC on catchment hydrology in large catchments: The Comet River, Central Queensland, Australia. J. Hydrol. 326: 199–214.

Skaggs, R.W., Amatya, D.M., Chescheir, G.M. and Blanton, C.D. 2006. Soil property changes during loblolly pine production. Paper no. 06-8206, ASABE, St. Joseph, MI.

Skaggs, R.W., Chescheir, G.M., Fernandez, G.P., Amatya, D.M. and Diggs, J. 2011. Effects of land use on soil properties and hydrology of drained coastal plain watersheds. Trans. ASABE 54(4): 1357–1365.

Skaggs, R.W., Gilliam, J.W. and Evans, R.O. 1991. A computer simulation study of pocosin hydrology. Wetlands 11: 399–416.

Smith, R. and Scott, D. 1992. The effects of afforestation on low flows in various regions of South Africa. Water S.A. 18(3): 185–194.

Spilsbury, M.J. 2002. Forests, water and mixed land use mosaics in developing countries; Making research solutions effective. pp. 266–274. *In*: Proc. of the Meeting on International Expert Meeting on Forests and Water, 20–22 November 2002, Shiga, Japan. International Forestry Cooperation Agency, Ministry of Agriculture, Forestry and Fisheries, Tokyo, Japan.

Ssegane, H., Amatya, D.M., Chescheir, G.M., Skaggs, R.W., Tollner, E.W. and Nettles, J.E. 2013. Consistency of hydrologic relationships of a paired watershed approach. Amer. J. Climate Change (In press).

Stednick, J.D. 1996. Monitoring the effects of timber harvest on annual water yield. J. Hydrol. 176: 79–95.

Swank, W.T. and Crossley, D.A. 1988. Forest Hydrology and Ecology at Coweeta. Springer-Verlag, New York, NY.

Swank, W.T. and Douglass, J.E. 1974. Streamflow greatly reduced by converting deciduous hardwood stands to pine. Science 185: 857–859.

Swank, W.T., Vose, J.M. and Elliott, K.J. 2001. Long-term hydrologic and water quality responses following commercial clearcutting of mixed hardwoods on a southern Appalachian catchment. Forest Ecology and Management 143(1): 163–178.

Sun, G., Alstad, K., Chen, J., Chen, S., Ford, C.R., Lin, G., Liu, C., Lu, N., McNulty, S.G., Miao, H., Noormets, A., Vose, J.M., Wilske, B., Zeppel, M., Zhang, Y. and Zhang, Z. 2011a. A general predictive model for estimating monthly ecosystem ET. Ecohydrology 4: 245–255.

Sun, G., Caldwell, P.V., Georgakakos, A.P., Arumugam, S., Cruise, J., McNider, R.T., Terando, A., Conrads, P.A., Feldt, J., Misra, V., Romolo, L., Rasmussen, T.C., McNulty, S.G. and Marion, D.A. 2013. Impacts of climate change and variability on water resources in the Southeastern U.S. pp. 210–236. *In*: Ingram, K.T., Dow, K. and Carter L. (eds.). Southeastern Regional Technical Report to the National Climate Change Assessment. Island Press.

Sun, G., Caldwell, P., Noormets, A., Cohen, E., McNulty, S.G., Treasure, E., Domec, J.-C., Mu, Q., Xiao, J., John, R. and Chen, J. 2011b. Upscaling key ecosystem functions across the conterminous United States by a water-centric ecosystem model. Washington, DC: Island Press. J. Geophys. Res. G00J05, doi:10.1029/2010JG001573.

Sun, G. and Lockaby, B.G. 2012. Water quantity and quality at the urban-rural interface. pp. 26–45. *In*: Laband, D.N., Lockaby, B.G. and Zipperer, W. (eds.). Urban-Rural Interfaces: Linking People and Nature. American Society of Agronomy, Crop Science Society of America, Soil Science Society of America, Madison, WI.

Sun, G., McNulty, S.G., Lu, J., Amatya, D.M., Liang, Y. and Kolka, R.K. 2005. Regional annual water yield from forest lands and its response to potential deforestation across the southeastern United States. J. Hydro. 308: 258–268.

Sun, G., McNulty, S.G., Moore Myers, J.A. and Cohen, E.C. 2008. Impacts of multiple stresses on water demand and supply across the Southeastern United States. J. Am. Water Resour. Assoc. 44(6): 1441–1457.

Sun, G., McNulty, S.G., Shepard, J.P., Amatya, D.M., Riekerk, H., Comerford, N.B., Skaggs, W. and Swift, L., Jr. 2001. Effects of timber management on the hydrology of wetland forests in the southern United States. For. Ecol. Manage. 143(1-3): 227–236.

Sun, G., Riekerk, H. and Kornhak, L.V. 2000. Ground-water-table rise after forest harvesting on cypress-pine flatwoods in Florida. Wetlands 20(1): 101–112.

Sun, G., Zhou, G., Zhang, Z., Wei, X., McNulty, S.G. and Vose, J.M. 2006. Potential water yield reduction due to forestation across China. J. Hydro. 328(3): 548–558.

Tang, Z., Engel, B.A., Pijanowski, B.C. and Lim, K.J. 2005. Forecasting LUC and its environmental impact at a watershed scale. J. Environ. Manage. 76: 35–45.

Thanapakpawin, P., Richey, J., Thomas, D., Rodda, S., Campbell, B. and Logsdon, M. 2007. Effects of landuse change on the hydrologic regime of the Mae Chaem river basin, NW Thailand. J. Hydro. 334(1-2): 215–230.

Tian, S., Youssef, M.A., Amatya, D.M. and Vance, E. 2013b. A global sensitivity analysis of DRAINMOD-FOREST, an integrated forest ecosystem model. Online in Wiley online library, Hydrological Processes, 2013, DOI:10.1002/hyp.9948.

Tian S., Youssef, M.A., Skaggs, R.W., Amatya, D.M. and Chescheir, G.M. 2012. DRAINMOD-FOREST: Integrated modeling of hydrology, soil carbon and nitrogen dynamics, and plant growth for drained Forests. J. Environ. Qual. 41: 764–782.

Troendle, C.A. and King, R.M. 1985. The fool creek watershed, 30 years later. Water Resour. Res. 21(12): 1915–1922.
Turner, K.M. 1991. Annual ET of Native Vegetation in a Mediterranean-Type Climate. Water Resources Bulletin 27(1): 1–6.
U.N. General Assembly. 2004. 59th Session. Accessed 27 February 2013. Prog. 11 Environment. http://www.un.org/ga/59/documentation/list0.html.
USDA. 2001. Forests: The potential consequences of climate variability and change. A report of the National Forest Assessment Group for the U.S. Global Change Research program. USDA Global Change Program Office, Washington, D.C.
U.S. Forest Service. 2011. Forest Service unveils first comprehensive forecast on southern forests. Asheville, N.C.: U.S. Forest Service Southern Research Station. Available at: www.srs.fs.usda.gov/news/472.
U.S. Forest Service. 2013. Riparian Forest Buffers: Function and Design for Protection and Enhancement of Water Resources. Radnor, PA. NA-PR-07-91. Accessed 27 February 2013. http://na.fs.fed.us/spfo/pubs/n_resource/buffer/cover.htm.
Van Wyk, D. 1987. Some effects of afforestation on streamflow in the Western Cape Province, South Africa. Water SA 13(1): 31–36.
Veraverbeke, S., Lhermitte, S., Verstraeten, W.W. and Goossens, R. 2011. A time-integrated MODIS burn severity assessment using the multi-temporal differenced normalized burn ratio (dNBRMT). Int. J. Appl. Earth Obs. Geoinf. 13(1): 52–58.
Verbesselt, J., Hyndman, R., Newnham, G. and Culvenor, D. 2009. Detecting trend and seasonal changes in satellite image time series. Remote Sensing of Environment 114(1): 106–115.
Vose, J.M., Ford, C.R., Lasyter, S., Dymond, S., Sun, G., Adams, M.B., Sebestyen, S., Campbell, J., Luce, C., Amatya, D., Elder, K. and Heartsill-Scalley, T. 2012. Can forest watershed management mitigate climate change effects on water resources? Revisiting Experimental Catchment Studies in Forest Hydrology (Proceedings of a Workshop held during the XXV IUGG General Assembly in Melbourne, June–July 2011) (IAHS Publ. 353, 2012) pp. 12–25.
Vose, J., Sun, G., Ford, C.R., Bredemeier, M., Ostsuki, K., Wei, X., Zhang, Z. and Zhang, L. 2011. Forest ecohydrological research in the 21st century: what are the critical needs? Ecohydrology 4(2): 146–158.
von Stackelberg, N.O., Chescheir, G.M., Skaggs, R.W. and Amatya, D.M. 2007. Simulation of the hydrologic effect of afforestation in the Tacuarembo River basin, Uruguay. Trans. of the ASABE 50(2): 455–468.
Wang, X., Melesse, A.M. and Yang, W. 2006. Influences of potential evapotranspiration estimation methods on SWAT's hydrologic simulation in a northeastern Minnesota watershed. Trans. ASABE 49(6): 1755–1771.
Webb, A., Kathuria, A. and Turner, L. 2012. Longer-term changes in streamflow following logging and mixed species eucalypt forest regeneration: The Karuah experiment. J. Hydrol. 464–465.
Wilk, J., Andersson, L. and Plermkamon, V. 2001. Hydrological impacts of forest conversion to agriculture in a large river basin in northeast Thailand. Hydro. Proc. 15: 2729–2748.
Williams, A.P., Allen, C.D., Macalady, A.K., Griffin, D., Woodhouse, C.A., Meko, D.M., Swetnam, T.W., Rauscher, S.A., Seager, R., Grissino-Mayer, H.D., Dean, J.S., Cook, E.R., Gangodagamage, C., Cai, M. and McDowell, N.G. 2013. Temperature as a potent driver of regional forest drought stress and tree mortality. Nature Climate Change 3: 292–297.
Williams, C.A., Reichstein, M., Buchmann, N., Baldochhi, D., Beer, C., Schwalm, C., Wohlfahrt, G., Hasler, N., Bernhofer, C., Foken, T., Papale, D., Schymanski, S. and Schaefer, K. 2012. Climate and vegetation controls on the surface water balance: Synthesis of ET measured across a global network of flux towers. Water Resour. Res. Vol. 48.
World Resources Institute. 1996. World Resources, 1996–97. Oxford University Press, Oxford.
Worrall, F., Swank, W.T. and Burt, T.P. 2003. Changes in stream nitrate concentrations due to land management practices, ecological succession, and climate: Developing a systems approach to integrated catchment response. Water Resour. Res. 39(7): 1177.
WWC. 2009. Water at a Crossroads: Dialogue and Debate at the 5th World Water Forum, Istanbul, Turkey. World Water Council (WWC), Conseil Mondial De Leau, Marselle, France.
Zegre, N. 2008. The history and advancement of change detection methods in forest hydrology. 2008. In Society of American Foresters Water Resources Working Group Newsletter. Society of American Foresters, Bethesda, Md.
Zégre, N., Skaugset, A.E., Som, N.A., McDonnell, J.J. and Ganio, L.M. 2010. In lieu of the paired catchment approach: Hydrologic model change detection at the catchment scale. Water Resour. Res. Vol. 46(11). W11544, doi:10.1029/2009WR008601, 2010.

Zhang, L., Dawes, W.R. and Walker, G.R. 2001. Response of mean annual ET to vegetation changes at catchment scale. Water Res. Res. 37(3): 701–708.

Zhang, X., Jin, C., Guan, D., Wang, A., Wu, J. and Yuan, F. 2012. Long-term eddy covariance monitoring of ET and its environmental factors in a temperate mixed forest in Northeast China. J. Hydrol. Engg. 17(9): 965–974.

Zhang, Y., Guan, D., Jin, C., Wang, A., Wu, J. and Yuan, F. 2011. Analysis of impacts of climate variability and human activity on streamflow for a river basin in northeast China. J. Hydrol. 410: 239–247.

8

Uncertainty is the Key Challenge for Agricultural Water Resources Management under Climate Change in the BRICS

Mark Mulligan,[1,*] *Martin Keulertz,*[1] *Bhopal Pandeya,*[1] *Mai Kivelä,*[1] *Lineu Neiva Rodrigues*[2] and *Tony Allan*[1]

Introduction

In this paper we examine agriculture in the BRICS countries (Brazil, Russia, India, China and South Africa). These are developing or newly industrialised countries with large, rapidly growing economies and/or a significant influence on regional or global international affairs. We examine the scenarios for climate change in these regions and use the WaterWorld Policy Support System (v. 2.90, which is a sophisticated hydrological model, see Mulligan, 2012) to examine the likely impacts of an IPCC SRES climate scenario A2a on agricultural water use in major agricultural areas of these countries, for an ensemble of 17 General Circulation Models (GCMs). We discuss the context for agriculture in these countries and the likely trajectories for agricultural development. We then examine the BRICS export markets by volume and destination and highlight the virtual water trades associated with agriculture. We discuss the key characteristics of climate change scenarios for these regions. The BRICS show very

[1] Department of Geography, King's College London, Strand, London, WC2R 2LS.
[2] Brazilian Agricultural Research Corporation - Embrapa.
[a] Email: mark.mulligan@kcl.ac.uk
* Corresponding author

different patterns and directions of agricultural water resource impact as a result of climate change but there is also huge uncertainty in the climate futures projected for these regions and this will make climate-smart agriculture very difficult to manage. These changes will have effects nationally but also along the very complex global supply chains associated with these BRICS economies. Finally, we examine other threats to agriculture in the BRICS and place these within the context of climate change and conclude that climate uncertainty is a key challenge for agricultural water resources management under climate change in the BRICS.

The Global Context

The topic of climate change in BRICS is very significant. The big four (BRIC) are all of global significance with respect to the supply of or the demand for food, water or both. They already are, or could be very significant players in global food supply chains and are shaping to change global food regimes. However, climate change has to vie for attention in the political economy with demographic change and management of agricultural and water productivity. These have presented themselves as much bigger challenges than climate change in the past couple of centuries—especially the past half century. There has been extraordinary response and adaptation to these non-climate related changes. Can such effective adaptability be expected as a response to climate change? This rather depends on the scale, rate and predictability of climate change that the BRICS are facing.

Study Areas

The areas analysed in this chapter are described in Table 8.1, each is a square tile of 10 degrees latitude and longitude. Though these tiles include areas outside the BRICS countries, such areas may also supply water to the BRICS country of interest (for example the Himalaya in the India tile) and are thus important to include, though we cannot simulate an entire country nor the watersheds of all of the rivers that supply it. Thus we focus on 10-degree tiles used by *WaterWorld*. The tiles cover only part of the country but are considered as representative of the kinds of changes that will need to be managed by each nation.

These tiles represent the populations, land areas, cover of cropland and pasture shown in Table 8.2. In each case the percentage in brackets denotes the proportion of

Table 8.1. Areas of analysis within the BRICS.

Country	North	South	West	East	Areas
Brazil	–10	–20	–50	–40	Minas Gerais, Tocantins, Bahia, Brasilia
Russia	60	50	40	50	Saratov, Ulyanovsk, Penza, Cheboksary, Kirov
India	30	20	80	90	Odisha, Chattisgarh, West Bengal, Jharkhand, Bihar, Uttar Pradesh
China	30	20	110	120	Fujian, Jiangxi, Guangdong, Guangxi, Hunan
South Africa	–30	–40	20	30	Mthatha, Queenstown, East London, Beaufort West

Table 8.2. Key characteristics of the study tiles (tile) in relation to the statistics for their national territories (nat) in solute values and by percent.

Country	Population (approx. persons) 2007	Land area (sq. km)	Cropland (sq. km)	Pasture (sq. km)
Brazil (nat)	26,000,000 (14%)	1,400,000 (14%)	99,000 (16%)	540,000 (25%)
Brazil (tile)	189,276,780	10,187,845	623,945	2,162,604
Russia (nat)	22,000,000 (16%)	1,400,000 (3%)	540,000 (22%)	190,000 (13%)
Russia (tile)	141,193,660	42,033,434	2,439,969	1,463,255
India (nat)	480,000,000 (43%)	1,400,000 (37%)	610,000 (29%)	130,000 (94%)
India (tile)	1,117,329,900	3,771,446	2,133,643	138,826
China (nat)	270,000,000 (20%)	1,000,000 (7%)	180,000 (9%)	71,000 (1.8%)
China (tile)	1,340,098,400	13,698,147	2,071,914	4,014,566
South Africa (nat)	7,500,000 (17%)	520,000 (32%)	33,000 (17%)	370,000 (38%)
South Africa (tile)	44,023,872	1,627,491	198,776	971,063

the total for the national territory that is within the tile. The tiles represent significant populations (480 million for India, 270 million for China) and between 14% (Brazil) and 43% (India) of the total national population. These populations represent significant markets for agricultural production within the case study tiles and within the BRIC countries even before considering agricultural exports from them. The tiles represent between 3% (Russia) and 37% (India) of national territory and contain from 9% (China) to 29% (India) of national total cropland area and 1.8% (China) to 94% (India) of total pasture area. These tiles are thus significant samples of the respective countries and thus any impacts of climate change within these areas will be significant at the national scale.

The Agricultural Context and Key Markets

It is important to understand that each of these countries has significant and heterogeneous international markets for agricultural commodities. We used the UN Comtradedatabase (2013) of export values (USD) to map the key flows of agricultural products from the BRICS as a means of highlighting international dependencies on agriculture in these countries. We calculate the proportion of national scale exports that could derive from the *WaterWorld* tile in question according to the distribution of associated land (on the basis of the percent of national territory within the tile for COMTRADE's 'all exports'; using Ramankutty et al. (2008) [Cropland] for COMTRADE 'crops'; Monfreda et al. (2008) [Vegetables and Melons] for COMTRADE 'vegetables, fruits and nuts'; and Ramankutty et al. (2008) [Pasture] for COMTRADE 'animal products'). The export values are re-scaled according to the proportion of the national total of each of these land covers that occurs within the tile. Thus the figures presented here are the dollar totals for the tile, not the country and this is important since the climate and water resources analysis here are also for the tile, not the country. This analysis helps in understanding the 'market' countries that are at risk from changes in food import commodity prices and availability that may result from climate change (or other changes) within the BRICS.

The summary of exports by BRICS country in Table 8.3 shows the China and Russia tiles to have the greatest total exports followed by Brazil, India and South Africa. For crops, the Brazilian tile exports are an order of magnitude greater than India and China which are in turn an order of magnitude greater than Russia and South Africa. For animal products, the Brazilian tile also has the highest export value followed by China, South Africa, India and Russia. India (at the national level) exports to the most countries followed by China, Brazil, South Africa and Russia. For crops India (at the national level) exports to the most countries followed by Brazil, China, South Africa and Russia and for animal products Brazil (again at the national level) exports to the most countries followed by South Africa, China, India and Russia. In terms of likely impacts on international agricultural trade, Brazil and India are thus the most significant of the BRICS in terms of total exports and also number of countries exporting to.

We now examine the destination of exports from the BRICS country tiles assuming that the destinations for products produced in the tile scale are the same as those produced at the national scale and documented by UN COMTRADE. This analysis serves to highlight the networks of food dependency between the BRICS and other countries.

Table 8.3. Exports by BRIC country, for the tile not for entire country. Analysed using inputs from UNComtrade, DESA/UNSD.

Country	Average annual total exports (for tile) 2007–2011 BN USD	Average annual crop exports (for tile) 2007–2011 BN USD	Average annual animal product exports (for tile) 2007–2011 BN USD	Number of countries exporting to (all exports, crops, animal products)
Brazil	69	0.94	0.09	188, 129, 102
Russia	210	0.001	0.003	161, 37, 34
India	30	0.11	0.009	200, 131, 74
China	170	0.1	0.02	199, 125, 82
South Africa	4.28	0.004	0.03	184, 94, 91

Brazil—the global "water tower"

Brazil is well known for its crop and livestock production miracle that converted the agriculturally poor expanse of cerrado into a highly productive agricultural landscape through the development of crop varieties combining tolerance to aluminum toxicity with high yield, better resistance to major diseases, and improved management practices such as liming and fertilizing to restore nutrients as well as crop rotation. Brazil is often considered one of the the world's breadbaskets and has long been the leading exporter of the world's breakfast foods (orange juice, coffee, sugar and cocoa). Latin America's largest economy is now also a leading producer of grains and meat (the world's largest beef herd), sugar (half the world's sugar market), corn (the third largest global exporter), soybeans (second only to the USA), timber and pulp, cotton (4th largest producer globally), tobacco (world's largest producer) and ethanol (world's largest exporter). From zero in 2010 Brazil is now a major exporter of corn for food, feed and biofuel. Indeed, Brazil provided a lifeline for US cattle farmers when US crops failed during the droughts of 2011. Though Brazil has ample rainfall at the national scale the cerrado has a semi-humid tropical savanna climate with a very strong dry season from April–September and significant irrigation. Brazil is a highly globalised, export heavy economy with total exports (all COMTRADE categories) to almost all countries globally (Fig. 8.1). Crop exports are significant to USA, other Latin American, European (e.g., Germany), African and South Asian countries as well as Russia. Vegetables and animal products show a similar pattern.

Russia—an Eastern "comeback"

Russia's agricultural revival started in the first decade of the current century. After the break-up of the Soviet Union in 1991, government subsidies were removed leading to a difficult decade of restructuring of the agricultural sector from a command-driven state sector to a market-driven economy. In particular, the livestock industry's outputs sharply declined in the 1990s pulling crop yields downwards too (Bokusheva and Hockmann, 2005; USDA, 2013). However since the 2000s and especially fuelled by a new agricultural subsidisation policy of the Duma in 2006, data on Russian grain yields indicate a strong rebound accompanied by private investment in Western Russia's "black earth" region where some of the best soils in the world for grain production are located (Campanale, 2011). Russia has become an increasingly significant exporter of wheat, barley and maize. A legacy of the communist past, farm sizes in Russia are generally large with 75 percent of producers farming on 5000 hectares or more (USDA, 2013). Russia has total exports to many countries in the region and the rest of the world, crop exports to Asia and Europe are particularly high as are vegetable exports (Fig. 8.2). The water-stressed Middle East is increasingly becoming an important market for cereal and livestock exports. Animal product exports to USA are also high. However, Russia's agricultural sector is still subject to severe investment shortfalls. In particular, heavy machinery is in poor condition due to a lack of public investment. Private investment in agriculture is still inhibited by weak institutions especially in the eastern part of the country (Bokusheva and Hockmann, 2005). As a result, Russia still scores last of the BRICS countries' in crop and animal product exports despite its growing opportunities in cereal and livestock trade.

India—the stressed traditionalist

Agriculture is a dominant sector in the Indian economy and is the source of livelihood for hundreds of millions of people across the country. Major river basins including the Indus, the Ganges and the Brahmaputra rivers have a long tradition of intensive agriculture and the diverse hydro-climatic conditions have supported a diversity of farming and agricultural systems. The major agricultural systems include rain-fed, subtropical, intensive temperate and rice systems (FAO, 2011). These are supported by large contiguous surface irrigation systems based on groundwater and an extensive

a

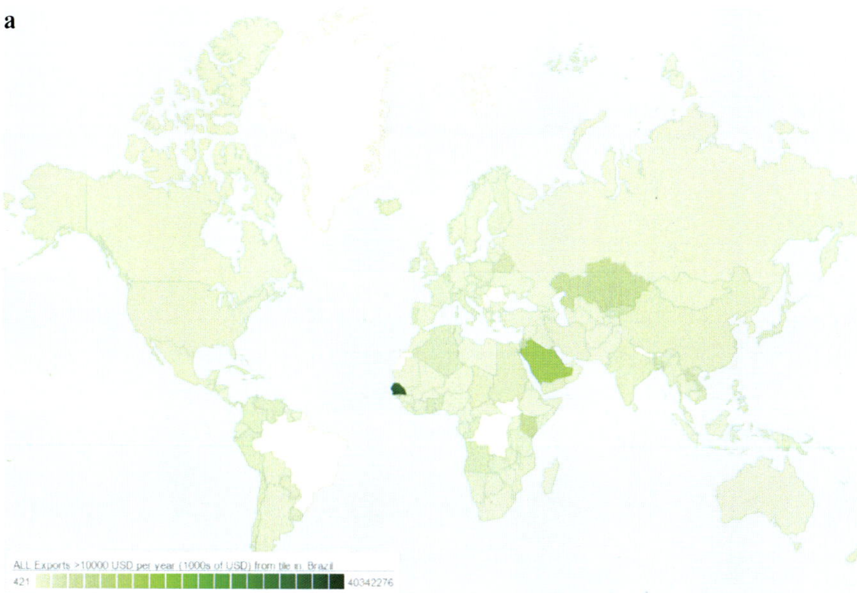

b

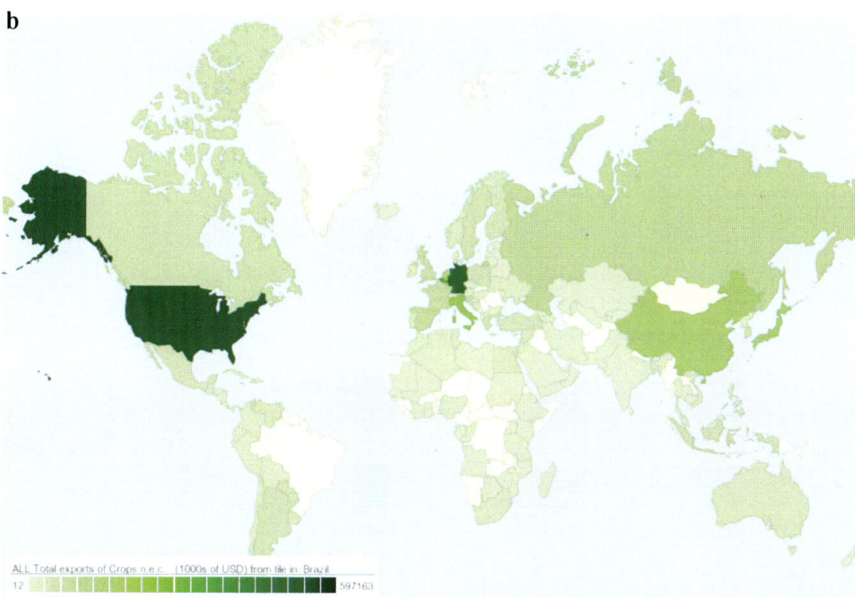

Figure 8.1. contd....

Figure 8.1. contd.

c

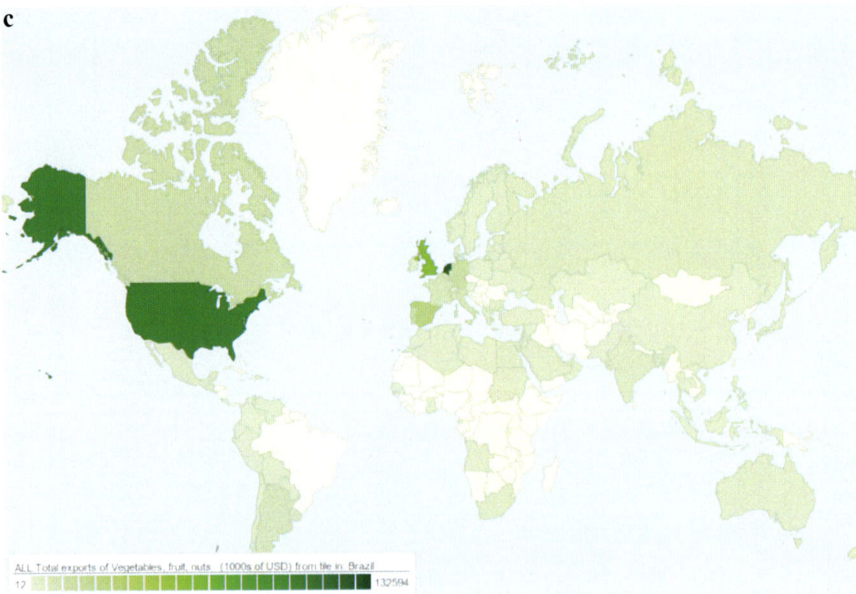

d

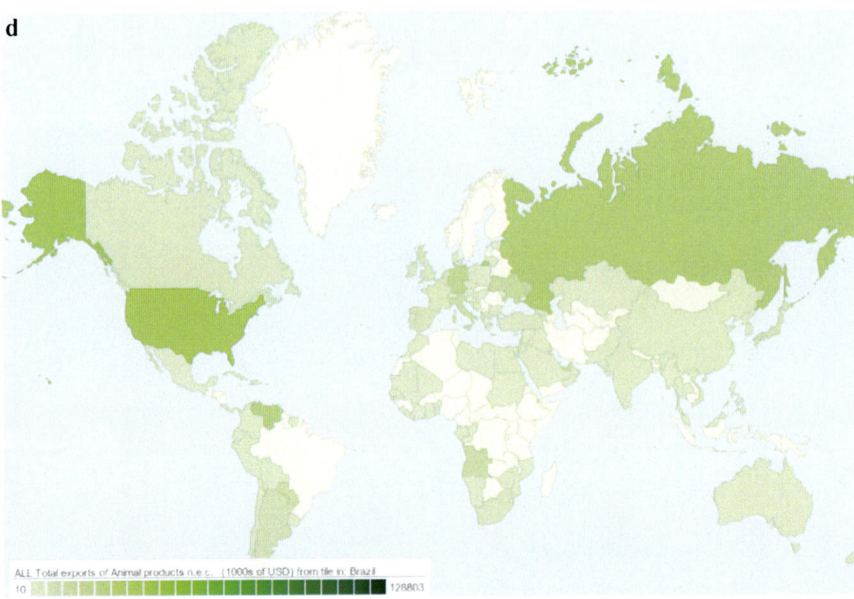

Figure 8.1. Destinations of export value from Brazil. Based on data from UNComtrade, DESA/UNSD.

a

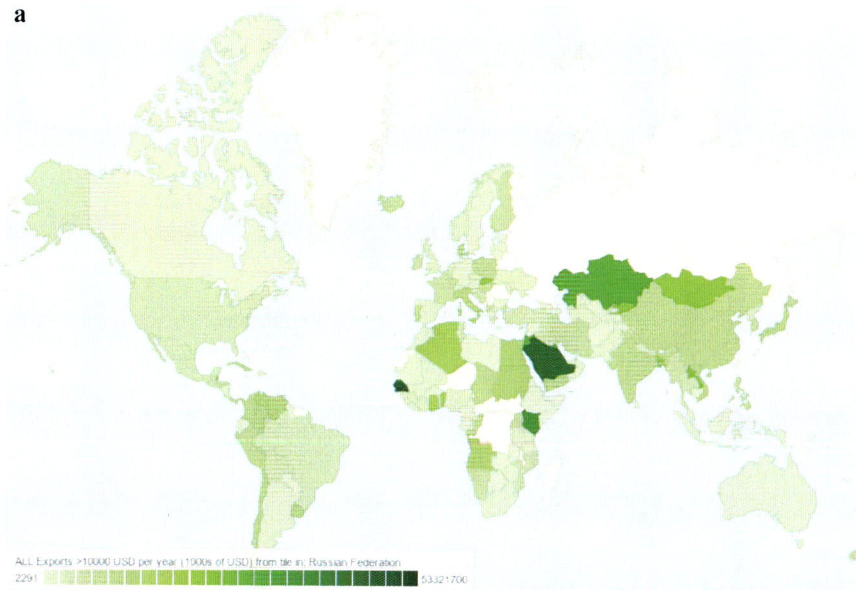

b

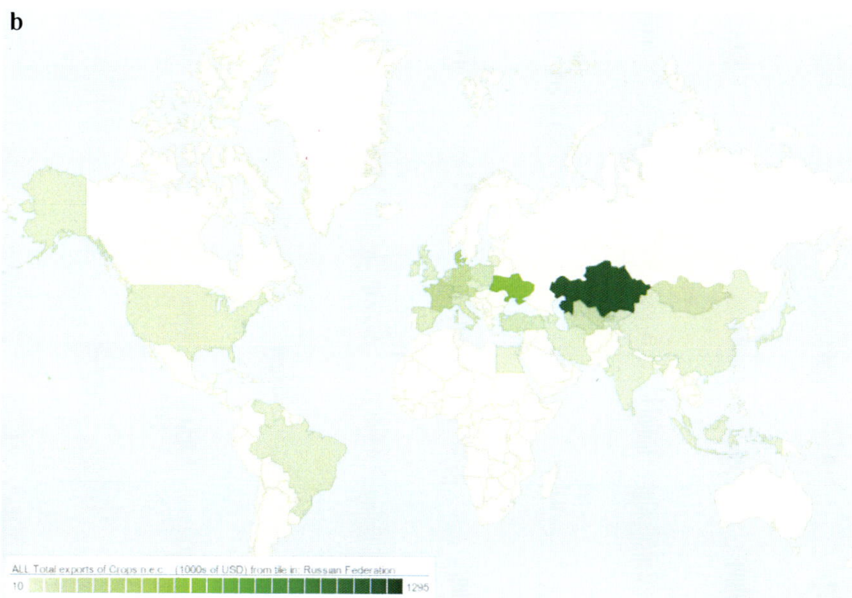

Figure 8.2. contd....

Figure 8.2. contd.

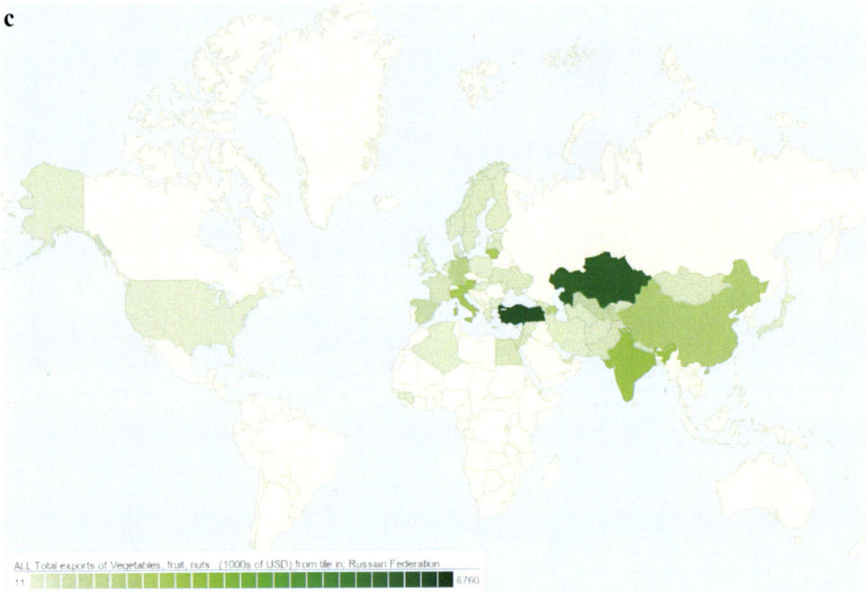

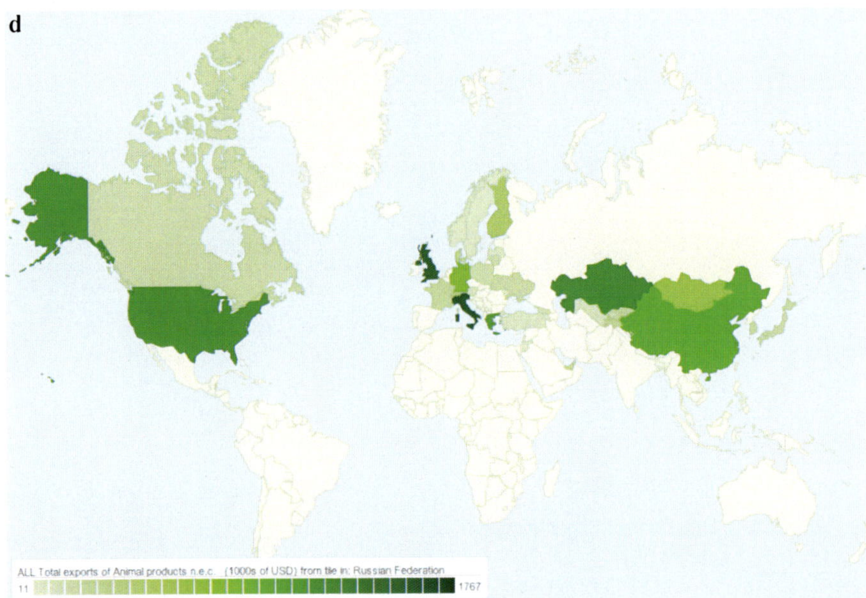

Figure 8.2. Destinations of export value from Russia. Based on data from UNComtrade, DESA/UNSD.

network of water storage dams. India has some of the world's largest irrigation schemes such as Bhakra, Gandak and Koshi Irrigation Systems. Over the last half century, the use of groundwater has been significantly increased and now it supplies about two thirds of the total irrigation water in India (Shah et al., 2006; Siebert et al., 2010). As a result, the nation's water resources are already hugely overstretched in parts of the country.

India is a global leader in agricultural production. The country has some 195 Mha of cropland of which 63 percent (125 Mha) is rainfed (World Bank, 2013). National food security depends on the production of sufficient cereal crops, as well as high value products such as fruits, vegetables, milk and meat to meet the demands of a growing population and changing food consumption patterns. There are several challenges to Indian agriculture. First, parts of the country are water stressed and water resources are unevenly distributed across the regions and the seasons. Secondly, increasing water productivity to meet demand requires significant investment. Thirdly, since the majority of poor people rely on agriculturally-based livelihoods, any climate impact on the agricultural production system will have direct consequences for the vast majority of these people. India exports to most countries, with crop exports particularly high to USA, Europe and Russia. Vegetables and animal products are exported to fewer countries (Fig. 8.3). India is the world's largest producer of milk, pulses, jute and spices; second in rice, wheat, sugarcane, farmed fish, sheep and goat meat, some vegetables, tea, fruits and cotton; and has the largest area under wheat, rice and cotton production worldwide (World Bank, 2013; GOI, 2013).

China—the global champion

China's agricultural sector employs approximately 300 million farmers making it a backbone of the rural economy. Chinese farmers produce more food than in any other country in the world even though good arable land is limited to the Eastern part of the country. However, the populous Chinese consumers at the same time have the highest total national consumption globally. As a result, the government has placed food self-sufficiency among the most important strategic policies in the country and has pressed farmers *not* to focus on cash crops for export markets. However, China has exports to many countries globally, with crop exports highest to Asia, Europe and USA and similar patterns for vegetables and animal products (Fig. 8.4).

South Africa—the "skilled parvenu"

South Africa's agricultural sector plays a "growth-permissive" role in the economy's development (Greyling, 2012). South Africa exports significantly in Africa, the Middle East and Asia with crop exports highest to Europe, China and the USA with similar patterns for vegetables and animal products (Fig. 8.5). The high exports enable South Africa to generate foreign currency to import other crops from other parts of the world for supply to domestic markets an affordable price. However, the farming sector suffers, like many sectors in the economy, from unresolved internal conflicts. As a result, a number of white farmers have moved further north into other African countries to utilise the perceived agricultural potential of the continent and also to extend South Africa's

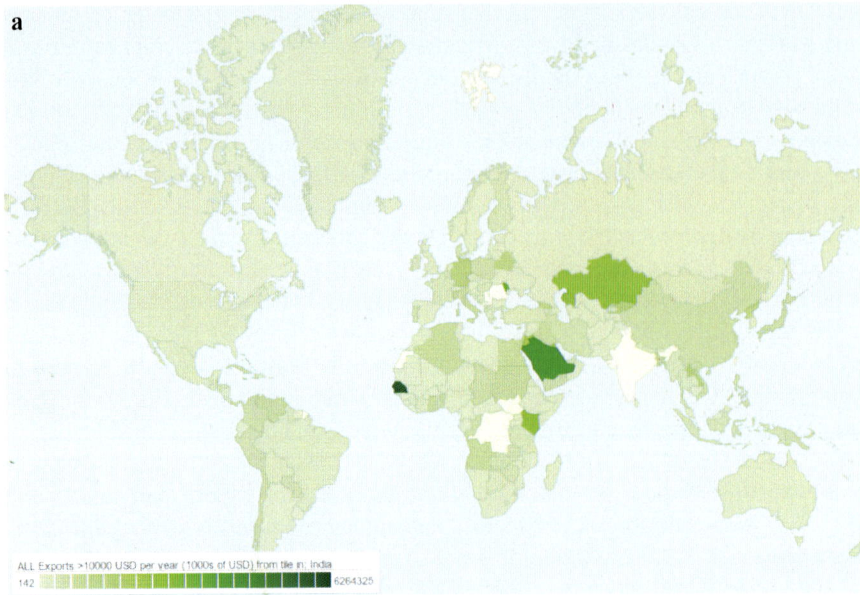

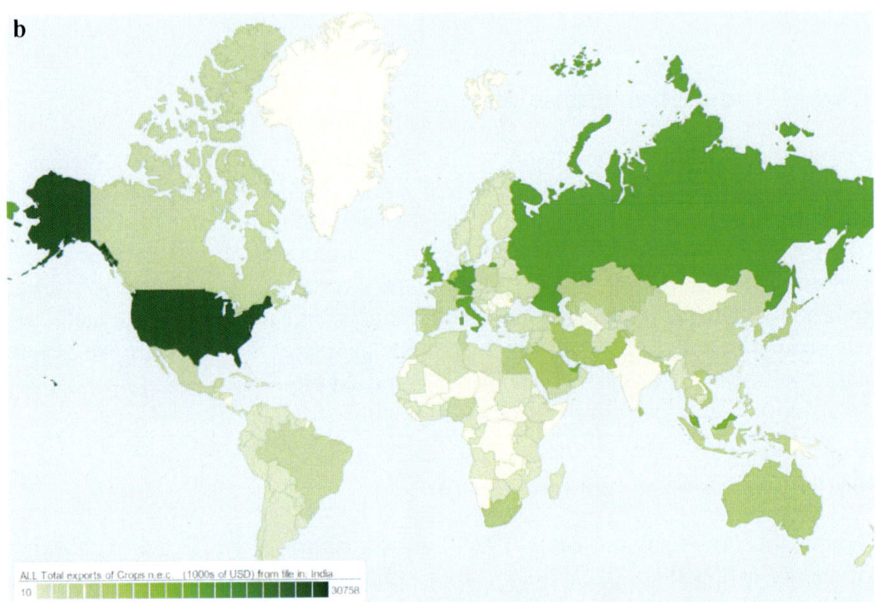

Figure 8.3. contd....

Figure 8.3. contd.

c

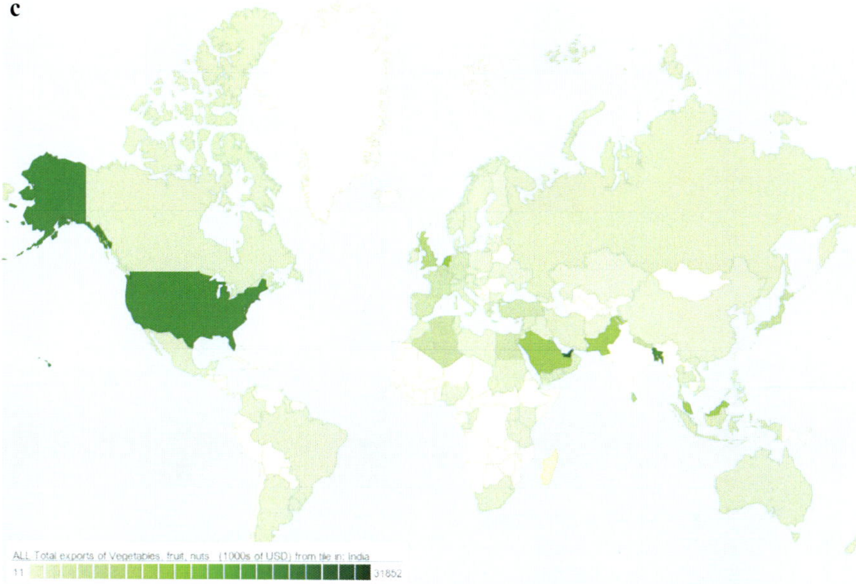

d

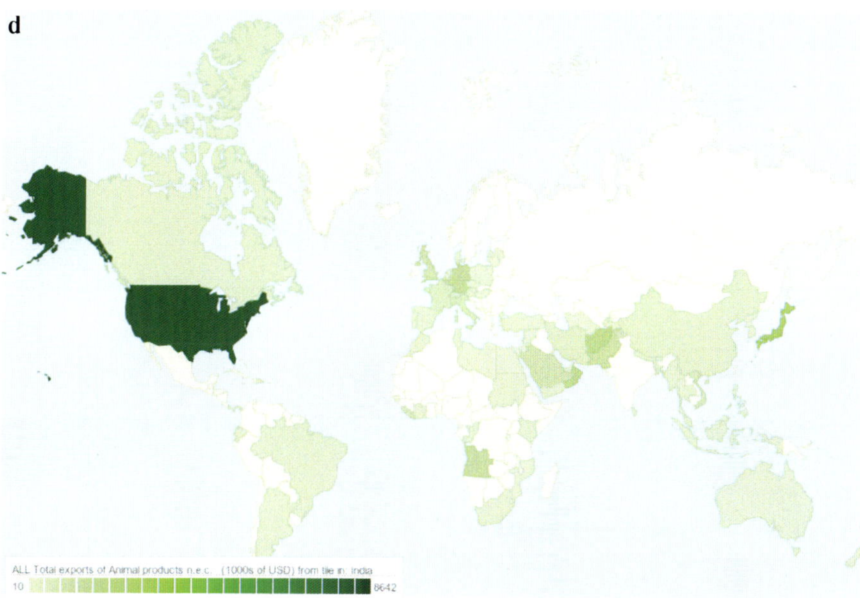

Figure 8.3. Destinations of export value from India. Based on data from UNComtrade, DESA/UNSD.

a

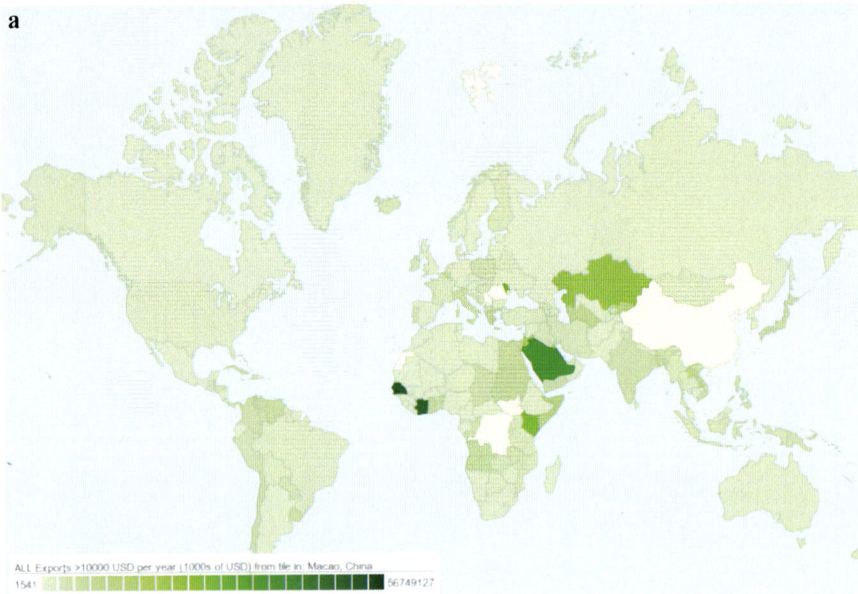

b

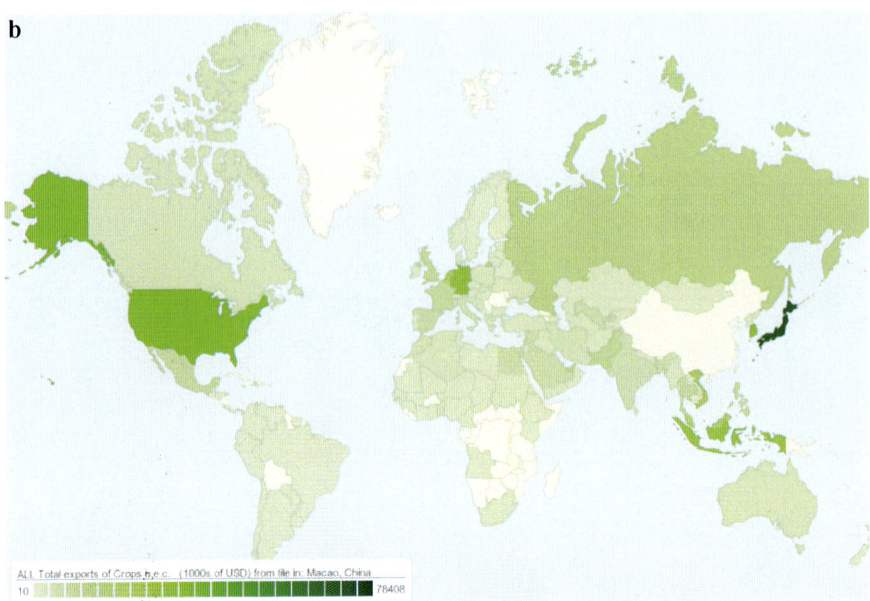

Figure 8.4. contd....

Figure 8.4. contd.

c

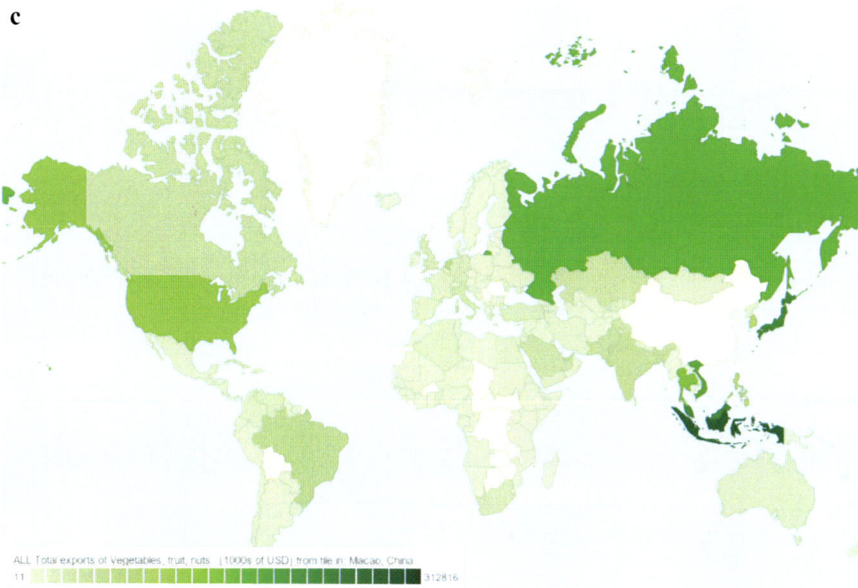

d

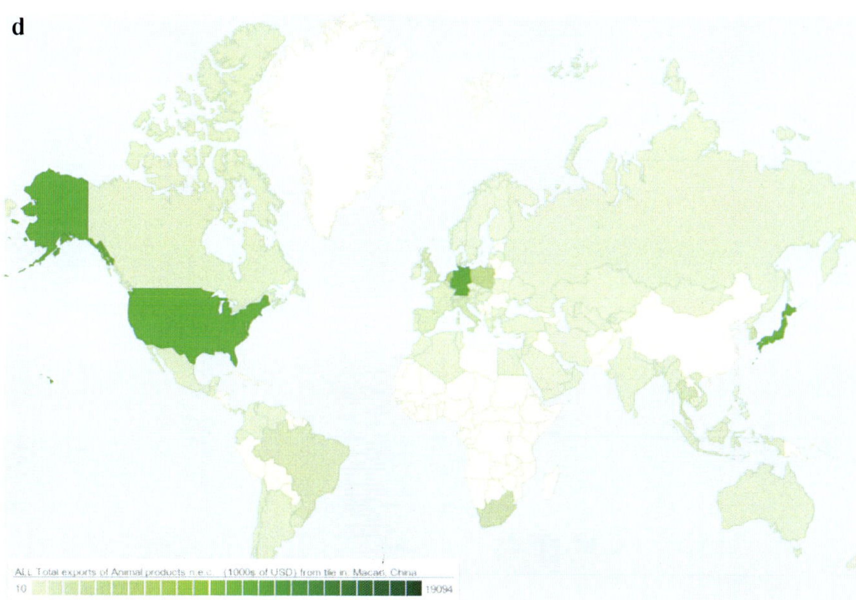

Figure 8.4. Destinations of export value from China. Based on data from UNComtrade, DESA/UNSD.

a

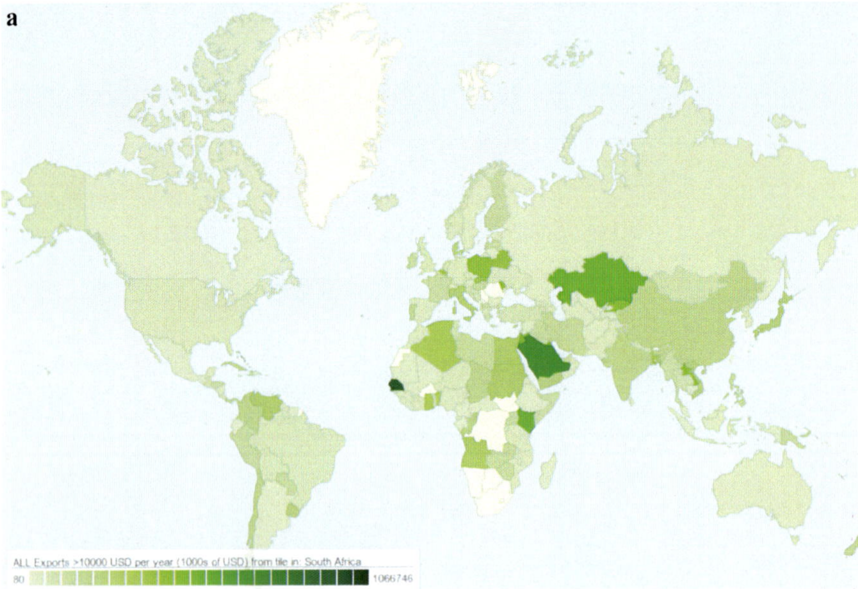

b

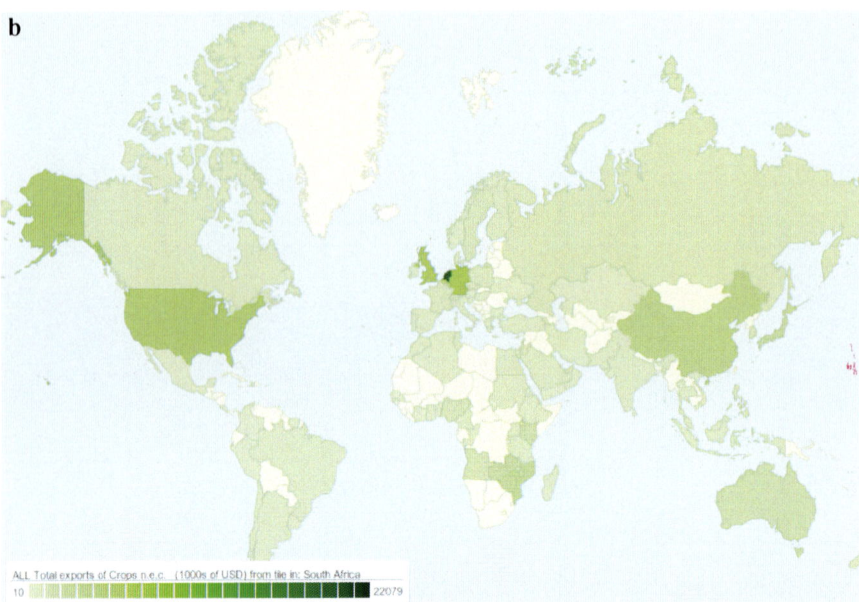

Figure 8.5. contd....

Figure 8.5. contd.

c

d

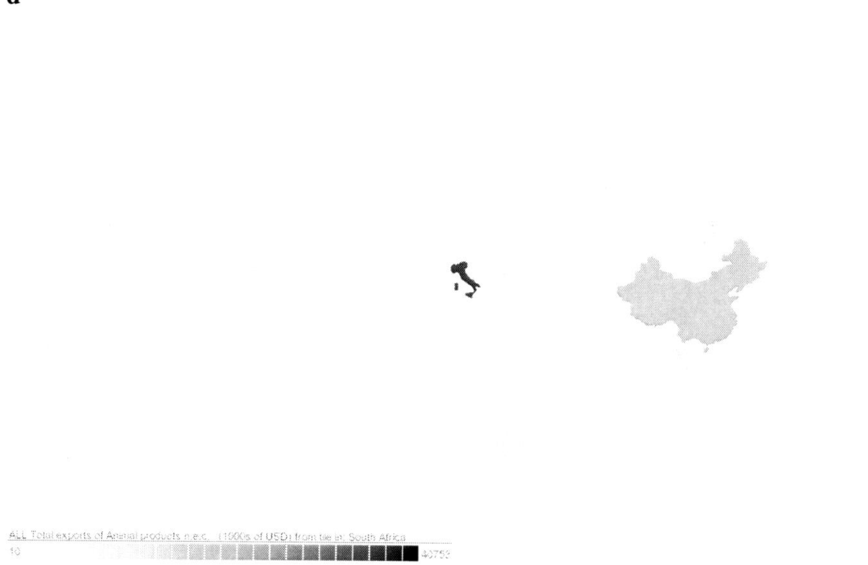

Figure 8.5. Destinations of export value from South Africa. Based on data from UNComtrade, DESA/UNSD.

sphere of influence across Africa (Pearce, 2011). South Africa's agricultural sector is thus increasing in strategic importance for the economy's geopolitical ambitions (Warner et al., 2013). Hence, South Africa will most likely determine the outcome of a wider agricultural development across Africa.

In conclusion the BRICS countries export crops and animal products globally. The connections between economy, agriculture and food security both within the BRICS and globally through these supply chains represent complex webs of global dependency on the climate and agricultural productivity of the BRICS. The mapped agricultural commodity flows embed significant exports of virtual water from the BRICS as discussed further in the next section.

Virtual Water Trade

Here we use the water footprint network data (Mekonnen and Hoekstra, 2011) to examine the virtual water imports and exports embedded in food and fibre trade with the BRICS. We calculate agricultural virtual water trade by summing the available metrics for virtual water flows related to trade in crop and animal products using both the import and export figures. Only the "blue" and "green" water footprint associated with production has been included in our analysis, not the grey water footprint associated with food processing and on-farm pollution. In the database 'flows' of virtual water have been calculated by multiplying the volume of trade by the respective average water footprint per ton of product in the exporting nation. Note that these water footprints are for both rainfed and irrigated agriculture and since they take no account of what land cover would exist in the absence of agriculture, may be misleading in cases where the natural vegetation would lead to equal or greater consumptive water use than the crop which replaces it. The Mekonnen and Hoekstra (2011) database covers 146 crops and over two hundred derived crop products as well as 8 farm animals (beef cattle, dairy cattle, pig, sheep, goat, broiler chicken, layer chicken and horses) and different animal products.

We see significant water exports from Brazil, India and China with China also importing significant volumes (Fig. 8.6). The global sum of international agricultural virtual water flows for the period 1996–2005 was 1,864,555 Mm^3/yr for import and 1,864,555 Mm^3/yr for export. This means that the BRICS countries constitute 12% of total import and 17% of total export of global virtual water trade of agricultural products. In the global ranking Brazil is 16th, Russia 10th, India 18th, China 2nd and South Africa 35th biggest importer of agricultural virtual water. On the export side Brazil is 2nd, Russia 18th, India 3rd, China 7th and South Africa 33rd largest exporter. These figures highlight the international dependencies of agricultural trade and show the importance of the BRICS countries, especially Brazil and India as significant net agricultural virtual water exporters. In the case of Brazil much of the agriculture replaces forest or savanna which may have equal or greater consumptive use of water to agriculture whereas for India much of the agriculture replaces sparsely vegetated drylands and is supported by significant irrigation. Conversion to agriculture in Brazil will thus have less of a water footprint on downstream available water quantity (relative to the prior natural cover) than conversion to agriculture in India, even for the same crops.

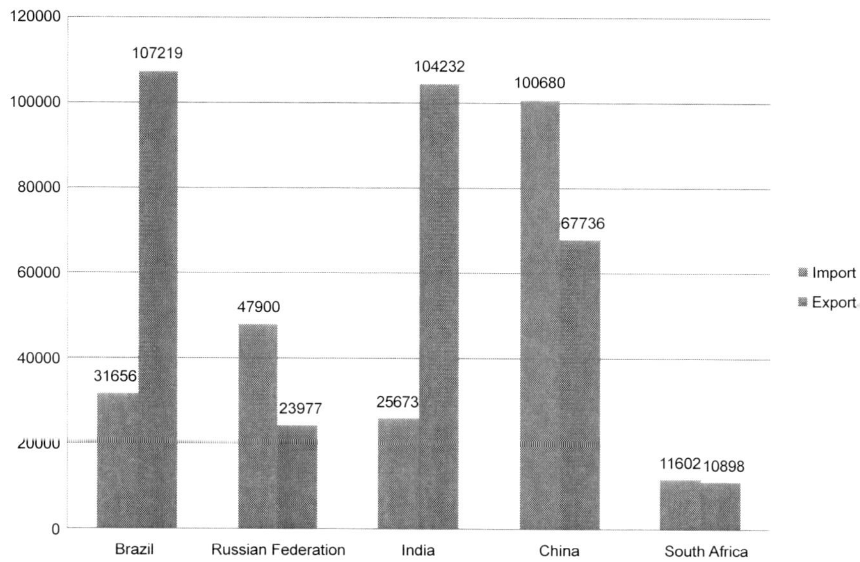

Figure 8.6. Virtual-water flows related to trade in food and fibre products (Mm^3/yr) (1996–2005). Source: based on data from Mekonnen and Hoekstra (2011).

Climate and Climate Change Scenarios for the BRICS

We used the WaterWorld Policy Support System v. 2.90 (Mulligan, 2012, www.policysupport.org/waterworld) to produce a *spatial ensemble mean* for the IPCC SRES A2a scenario (2050s) monthly temperature and precipitation outputs for key agricultural areas in each of the BRICS. The *spatial ensemble mean* combines the result of all 17 available GCMs and calculates the mean change in monthly and annual temperature and precipitation across all GCMs for each pixel in a 1 arc-degree tile at 1 km pixel resolution. This attempts to control for the significant differences in temperature and—especially—precipitation projections between GCMs and provide a consensus projection as the ensemble mean (Tebaldi and Knutti, 2010). The scenarios are applied in equilibrium mode by comparing the baseline climate (1950–2000) with the scenario climate (2050s).

Baseline climate

For each tile, statistics of the baseline climate (1950–2000, from WorldClim, Hijmans et al., 2005) were produced. These use raw WorldClim (Hijmans et al., 2005) temperature data and WorldClim rainfall corrected by WaterWorld for wind-driven effects (see Mulligan and Burke, 2005; Mulligan, 2012). The following paragraphs describe the baseline climate for each country.

Brazil—warmer, wetter, reduced growing season

For the baseline (1950–2000), Brazil shows an area-average precipitation of 1,187 mm/yr. Seasonally, precipitation for the area has an area-average maximum of 220 mm/month in December and 200 mm/month in January and a minimum of 9.6 mm/month in August and 9.9 mm/month in July. Annual mean area-average temperature is 23.3°C with an area-average maximum of 27.5°C and a minimum of 15.5°C. Area-average growing season length (defined as the number of months with a positive water balance and a mean monthly temperature > 6°C) is 7 months.

The ensemble mean A2a 2050s scenario produces a small rise in area-average annual total precipitation to 1202 mm/yr with decreases over 28% of the area and increases over 72%. Seasonally, maximum area-average precipitation was stable at 220 mm/month in December and 200 mm/month in January and rose to an area-average annual minimum of 23 mm/month in May and 24 mm/month in July. Mean annual area-average temperature increases by almost 3° overall to 25.9°. Growing season length declines over 34% of the area and increases over 14% of the area with the area-average mean falling to 6.8 months.

Russia—warmer, wetter, no change in growing season

For the baseline (1950–2000), Russia shows an area-average precipitation of 370 mm/yr. Seasonally, precipitation for the area has an area-average maximum of 71 mm/month in July and 61 mm/month in June and a minimum of 0 mm/month in January and 0 mm/month in February. Annual mean area-average temperature is 4.3°C across the study area with a maximum of 7.5°C and a minimum of 1.1°C. Growing season length is on average 6.8 months.

The ensemble mean A2a 2050s scenario produces a rise in average precipitation to 390 mm/yr with decreases over 21% of the area and increases over 79%. Seasonally, maximum monthly area-average precipitation fell slightly to 68 mm/month in July and 60 mm/month in June and the area-average minimum was stable at 0 mm/month in January and 0 mm/month in February. Mean annual area-average temperature increases by 3.4°C overall to 7.7°C. Growing season length declines over 14% of the area and increases over 18% of the area but remains the same over 68% of the area with a mean area-average change of only –0.015 months.

India—warmer, wetter, increased growing season

For the baseline (1950–2000), India shows area-average precipitation of 1,200 mm/yr. Seasonally, precipitation for the area has a maximum of 320 mm/month in July and 300 mm/month in August and a minimum of 4.4 mm/month in December and 7.4 mm/month in November. Annual mean temperature is 20.3°C across the study area with an area-average maximum of 27.7°C and a minimum of –22°C. Growing season length is on average 5.3 months.

For the scenario, area-average precipitation rose slightly to 1,300 mm/yr with decreases over 1% and increases over 99% of the area. Seasonally, precipitation for the area rose to a maximum of 340 mm/month in July and 310 mm/month in August and rose to an area-average minimum of 16 mm/month in December and 17 mm/month in November. Mean annual area-average temperature increases by 3° overall to 23.3°. Growing season length declines over 2% of the area and increases over 43% of the area with a mean of 5.9 months, increasing slightly on the baseline.

China—warmer, reduced growing season

For the baseline (1950–2000), the China tile shows area-average precipitation of 1,575 mm/yr. Seasonally, precipitation for the area has an area-average maximum of 250 mm/month in June and 250 mm/month in May and a minimum of 41 mm/month in December and 51 mm/month in January. Annual mean area-average temperature is 18°C across the study area with a maximum of 24°C and a minimum of 0°C. Growing season length is on average 11.1 months.

For the mean A2a 2050s scenario, area-average precipitation rose slightly to 1585 mm/yr with decreases over 20% of the area and increases over 70%. Seasonally, area-average precipitation for the area rose slightly to a maximum of 260 mm/month in June and 250 mm/month in May and fell to a minimum of 37 mm/month in December and 41 mm/month in January. Mean annual area-average temperature increases by 2.5° overall to 20.5°. Growing season length declines over 24% of the area and increases over 5% of the area with the mean area-average value falling slightly to 10.8 months.

South Africa—warmer, drier, reduced growing season

For the baseline (1950–2000), South Africa shows an area-average precipitation of 440 mm/yr. Seasonally, precipitation for the area has an area-average maximum of 61 mm/month in March and 56 mm/month in February and a minimum of 17 mm/month in July and 18 mm/month in June. Annual mean area-average temperature is 15.7°C across the study area with a maximum of 20°C and a minimum of 0°C. Growing season length is on average 5.6 months.

For the scenario, area-average precipitation fell to 430 mm/yr with decreases over 60% of the area and increases over 40%. Seasonally, precipitation for the area remained stable at a area-average maximum of 60 mm/month in March and 57 mm/month in February and fell slightly to an area-average minimum of 16 mm/month in June and 17 mm/month in July. Mean annual area-average temperature increases by 2.4°C overall to 18.1°C. Growing season length declines over 31% of the area and increases over 5% of the area with an area-average decrease to a mean of 5.2 months.

Potential Impact of Climate Change on Agricultural Water Use

WaterWorld is a sophisticated water balance model capable of modelling the impacts of climate change scenarios on the baseline water balance for any region globally. Full details of the model specification and input datasets are given in Mulligan (2013). WaterWorld does not incorporate water stores such as groundwater but since—in the long term—available groundwater storage is a function of the water balance, these changes in water balance are as critical in the long term for areas in which agriculture is dominated by blue water (irrigated) as they are for green water (rainfed) inputs. WaterWorld was forced using baseline (WorldClim, 1950–2000) conditions and land use representative for 2010. It was then forced using the SRES A2a climate scenario downscaled to 1 km resolution using the delta method relative to the WorldClim climate baseline (see Mulligan et al., 2011). Only temperature and rainfall was changed with other climate variables (mean sea-level pressure, cloud frequency, relative humidity and wind speed) remaining at their baseline levels since these outputs are not provided by IPCC GCM outputs. Population, land use and other WaterWorld inputs also remained at their baseline levels. Key outputs of WaterWorld such as water balance were viewed as differences from the baseline value for the same variable. The croplands and pastures map of Ramankutty et al. (2008) was used to represent current croplands and pastures and the water balance output of WaterWorld masked by these maps for all croplands (defined as croplands >0 fractional) and pastures (defined as pastures >0 fractional) and for intensive croplands (defined as croplands >0.5 fractional) and pastures (defined as pastures >0.5 fractional). Results are described and summarised by country in the following paragraphs.

Brazil—reduced water and growing season overall but geographically variable

For the Brazil tile, the scenario led to a decrease in water balance over 7% of croplands (1.4 Mkm^2) and an increase over 29% with a mean area-average decline of 29 mm/yr. Pastures cover broadly the same area and thus the results are the same. For intensive croplands (6300 km^2) water balance decreased over 79% of areas and increased over 21% with a mean decline of 32 mm/yr. Intensive pastures cover some 470,000 km^2 of which 67% shows a decreased water balance and 33% an increase with an overall mean decrease of 35 mm/yr. For 31% of intensive pastures mean growing season length decreases (by around 1.6 months), 18% have an increase in growing season length (mean 1.5 months) and 51% show no change. For intensive croplands 55% show a decrease in mean growing season length of around 1.6 months with 37% showing no change and 8% showing an increase (1.3 months). In summary, the Brazil tile shows reduced suitability for rainfed agriculture and an increase in the irrigation demand under the ensemble mean scenario.

Russia—increased water overall and increased growing season

For the Russian tile, the ensemble mean scenario led to a decrease in water balance over 6% of croplands (1.3 M Mkm^2) and an increase over 94% with a mean increase of 55 mm/yr. Pastures cover the same area and show the same results. For intensive croplands (630,000 km^2) water balance decreased over 1% of areas and increased over 99% with a mean increase of 53 mm/yr. Intensive pastures cover some 11,000 km^2 of which 4% shows a decreased water balance and 96% an increase with an overall mean increase of 25 mm/yr. For 72% of intensive pastures mean growing season length increases (by around 1.7 months), 6% show a decrease of around 1.1 months and 22%

no change. For intensive croplands 14% show an increase in mean growing season length of around 1 month with 25% showing a decrease of around 1.5 months and 62% showing no change. The projected agricultural future for this region of Russia is thus projected to improve in terms of water resources and thermal growing season.

India—significantly more water and increased growing season

For India, the ensemble mean scenario led to a decrease in water balance over 1.7% of croplands (1.3 Mkm^2) and an increase over 98% with a mean area-average increase of 110 mm/yr. Pastures cover 1.1 Mkm^2 and show a decrease in water balance over 1.7% of areas and an increase over 98% with a mean increase of 110 mm/yr. For intensive croplands (660,000 km^2) water balance decreased over 0% of areas and increased over 100% with a mean increase of 110 mm/yr. Intensive pastures cover some 88,000 km^2 of which 0% shows a decreased water balance and 100% an increase with an overall mean increase of 110 mm/yr. For 91% of intensive pastures mean growing season length increases (by around 1.9 months) whilst for intensive croplands 36% show an increase in mean growing season length of around 1.3 months with 64% showing no change. The projected agricultural future for this region of India is thus positive, especially with respect to water resources supply.

China—less water but geographically variable impacts on growing season

For China, the ensemble mean scenario led to a decrease in water balance over 91% of croplands (which cover 960,000 km^2) and an increase over 9% with a mean decline across all areas of 66 mm/yr. Pastures cover 900,000 km^2 of which 91% showed a decrease in water balance and 9% an increase with a mean decline across all areas of 67 mm/yr. For intensive croplands (49,000 km^2) water balance decreased over 98% of areas and increased over 2% with a mean decline across all areas of 48 mm/yr. Intensive pastures cover some 9000 km^2 of which 91% shows a decreased water balance and 9% an increase with an overall mean decrease of 66 mm/yr. For intensive pastures 11% show an increase in mean growing season length of around 1 month with 53% showing no change and 36% a decrease of around 1.3 months. For intensive croplands 12% show an increase in mean growing season length of around 1 month with 68% showing no change and 20% a decrease of around 1.4 months. The projected agricultural future for the China tile is thus negative with respect to water resources resulting in a decrease in growing season length over some cropland and pastures but increases over others.

South Africa—less water and reduced growing season

For South Africa the ensemble mean scenario led to a decrease in water balance over 92% of croplands (450,000 Mkm^2) and an increase over 8% with a mean decline of 43 mm/yr. Pastures cover some 510,000 km^2 of which 89% show a decrease in water balance and 11% an increase. The mean decline over all areas is 39 mm/yr. For intensive croplands (5,500 km^2) water balance decreased over 100% of areas with a mean decline of 140 mm/yr. Intensive pastures cover some 410,000 km^2 of which 91% shows a decreased water balance and 9% an increase with an overall mean area-average decrease of 38 mm/yr. For 6% of intensive pastures mean growing season length increases (by around 1.2 months), but for 62% there is no change and for a further 32% a decrease of around 1.4 months. For intensive croplands 58% show a decrease in mean growing season length of around 1.2 months with 42% showing no change. The projected agricultural future for the South Africa tile is thus negative with respect to water resources.

Seasonality

So far we have looked mainly at annual impacts but climate change is also expected to affect water resources seasonality and in some areas this may have more significant agricultural impacts than the annual changes discussed. The seasonality index of Walsh and Lawler (1981) modified to handle negative values was calculated for the baseline water balance and for the water balance scenarios for comparison. Table 8.4 summarised the tile-average seasonality values which may mask much greater seasonality in some parts of the tile. An index value of greater than 0.4 is considered *seasonal*, >0.8 *marked seasonal* with a long dry season and >1.2 *extreme seasonal* with almost all water available in 1–2 months. In all cases the area-average water balance seasonality index increases with the application of the climate scenario. In some cases the increase is down to rainfall (e.g., China, South Africa) and in others it is down to increased seasonality in evapotranspiration (e.g., Brazil, Russia, India). Water balance is less seasonal than rainfall because of the impacts of evapotranspiration (ET) seasonality.

Table 8.4. Seasonality statistics for baseline and scenario for the BRIC tiles based on Walsh and Lawler (1981).

Country	Change in tile-average water balance seasonality	Change in tile-average rainfall seasonality	Change in tile-average ET seasonality
Brazil	baseline: 0.0067	baseline: 0.73	baseline: 0.192
	scenario: 0.019	scenario: 0.68	scenario: 0.195
Russia	baseline: 0.024	baseline: 0.82	baseline: 0.92
	scenario: 0.14	scenario: 0.66	scenario: 0.93
India	baseline: 0.0063	baseline: 1	baseline: 0.35
	scenario: 0.015	scenario: 0.92	scenario: 0.34
China	baseline: 0.0069	baseline: 0.49	baseline: 0.354
	scenario: 0.0094	scenario: 0.5	scenario: 0.352
South Africa	baseline: 0.128	baseline: 0.38	baseline: 0.536
	scenario: 0.133	scenario: 0.39	scenario: 0.544

Contributions to Change in Water Balance

Table 8.5 shows the contributions of different fluxes to the projected change in water balance. In some of the countries (e.g., India) the response is dominated by the

Table 8.5. Contribution of area-average change in different fluxes to the area-average change in water balance and overall outcome of the ensemble mean. Percentage in brackets are percentage change relative to the baseline for the flux of interest.

Country	Outcome	Change in water balance (mm/yr)	Change in rainfall (mm/yr)	Change in ET (mm/yr)	Change in fog inputs (mm/yr)	Change in snow and ice melt (mm/yr)
Brazil	negative	–29 (–17%)	+15 (+0.68%)	+40 (+5.2%)	–3.5 (–5.5%)	0
Russia	positive	+53 (+18%)	+26 (+6.4%)	+32 (+18%)	–1.3 (–11%)	+0.038 (+21%)
India	positive	+100 (+53%)	+150 (+19%)	+44 (+8%)	–1.6 (–11%)	–2.4 (–25%)
China	negative	–67 (–7.9%)	+11 (+0.54%)	+60 (+8.6%)	–15 (–65%)	0
South Africa	negative	–40 (–40%)	–13 (–0.48%)	+27 (+5.5%)	–0.04 (–4.1%)	0

rainfall change, in others, e.g., China and Brazil), the response is dominated by the temperature-driven change in actual evapotranspiration (ET). The change in ET is consistently an increase whereas the change in rainfall varies between regions from increases to decreases. Change in fog inputs are only mildly significant in the case of China and snow and ice melt as a result of climate change is only mildly significant in the case of India (though still an order of magnitude lower than the change in rainfall). Rainfall is thus a highly significant component of the change in water resources we can expect with climate change and an increase in rainfall with climate change is capable of converting a reduction of water resources that would normally result from increased evapotranspiration under climate warming to an increase in water resources.

Sensitivity to the Summary Metric for the GCM Ensemble

Though we have shown significant differences in annual and seasonal water balance and growing season both of which impact upon agricultural potential, these results are highly sensitive to the projected rainfall. GCM rainfall is highly uncertain so this sensitivity is critical. In order to better understand the sensitivity of these results to the ensemble mean which we assume to be a good representation of the 17 GCM ensemble, we run the same analysis with two other ensemble statistics. We forced WaterWorld with the temperature and rainfall ensemble mean plus one standard deviation of the per-pixel differences between the 17 GCM results (the mean+1SD, the 'hot, wet end') and then again for the mean minus one standard deviation of the differences between the 17 GCMs (the mean–1SD, the 'cool, dry end'). Table 8.6 shows the water balance

Table 8.6. The impact of GCM uncertainty on water balance outcomes (figures in square brackets denote the percentage of the area within which the same sign as the observed change is observed).

Country	Change in water balance (mean)	Change in water balance (mean – 1 SD)	Change in water balance (mean + 1 SD)
Brazil	–29 (–17%)	–260 (–92%) [100%]	+210(+74%) [99%]
Russia	+53(+18%)	–94(–34%) [99%]	+234(+81%) [100%]
India	+100(+53%)	–160(–62%) [99%]	+370(+130%) [100%]
China	–67(–7.9%)	–270(–31%) [100%]	+130(+15%) [99%]
South Africa	–40(–40%)	–190(–120%) [100%]	+110(+63) [97%]

results for the mean, the mean–1SD and the mean+1SD. In each case the difference in water balance between the mean–1SD and the mean+1SD is one of direction as well as magnitude of change. The mean–1SD shows drying for all regions (and 100% of the area) whereas the mean+1SD shows wetting for all regions. However, the ensemble mean shows drying for some (Brazil, China, South Africa) and wetting for others (Russia and India). This indicates that the uncertainty in water balance (largely driven by rainfall uncertainty) is so high that water resource futures are very difficult to project for these countries. If we see mean–1SD as the worst case agricultural water resources scenario then all countries dry, with Brazil, India and South Africa showing reductions of more than 50% over almost all of the studied areas. If, on the other hand, mean+1SD is the best case scenario then all countries show wetting relative to the baseline with greater than 100% wetting for all countries except China.

One could argue that a subset of the 17 models are the best predictors and only these should be used in impact analysis but since that 'best' subset of models varies geographically and there is no possibility for validation of projections, such a strategy is both impractical and risky on a regional scale, hence the use of the full ensemble here.

Projections for growing season length are a little more consistent (Table 8.7) with the mean and mean–1SD showing the same sign for all but Russia but the sign changing between mean and mean+1SD for all but India. Change is much less consistent over space within the tiles with often much less than 100% of the area showing the same sign of change (figures in square brackets) as the overall mean for the tile. Whilst growing season length is projected to decline for all tiles except India under the mean scenario, under the worst case scenario (mean–1SD), growing season for all countries declines by up to 2 months and under the best case scenario, growing season increases for all countries.

Even given a specific emissions scenario, we can see that the uncertainty between GCMs (especially for rainfall) renders them largely useless for adaptation planning at the significantly regional scale. Whilst we can get an idea of differences between the BRICS, these differences are highly sensitive to the GCM used. For now we can only say that climate will vary and change (it always has in the past) and its impacts on agricultural water at the regional to national scale in the BRICS will largely depend upon rainfall trends (though other factors may become important locally). We cannot predict the rainfall future so agriculture has to be able to adapt to a range of potential rainfall futures, whilst remaining robust to other non-climatic threats.

Table 8.7. The impact of GCM uncertainty on length of growing season outcomes (figures in square brackets denote the percentage of the area under which the same sign as the observed change is observed).

Country	Change in growing season length (mean)	Change in growing season length (mean – 1 SD)	Change in growing season length (mean + 1 SD)
Brazil	–0.26 (–4.2%) [34%]	–1.5(–23%) [77%]	+1.2 (+19%) [66%]
Russia	–0.015 (–1%)	–1.5 (–23%) [52%]	+1.4(25%) [81%]
India	+0.67(+16) [44%]	–0.33(–4.2%) [26%]	+1.6(+38%) [74%]
China	–0.29(–2.8%) [25%]	–2.4 (–22%) [89%]	+0.65 (+7%) [37%]
South Africa	–0.38(–7.7%) [31%]	–2.1 (–39%) [71%]	+1.2(+28%) [56%]

Non-climatic Threats

Markets and trade

As we have seen, climate change may cause negative outcomes for Brazil, China and South Africa while India and Russia may benefit. The increasing influence of all BRICS economies on global trade and markets may follow a Western agro-strategic pattern. The Western world has vertically and horizontally integrated its supply chains to increase efficiency and market power of food processors, traders and retailers since the middle of the twentieth century (Teubal, 1993). As a result, a corporate "food regime" has emerged with strategic nodes of power in global agriculture (McMichael, 2009; Friedmann, 1993; 1997 among others). For example, three American and one European agribusiness companies: Archers Daniels Midlands, Bunge, Cargill and Louis Dreyfus (the 'ABCDs') control 90% of global grain trade, fostered and aided by Western agricultural subsidies since the post-World War II period (Murphy et al., 2012). With the control of grain trade comes the control of global "virtual water" trade, which may pose an increasing problem to emerging economies such as the BRICS (Sojamo et al., 2012). As a result, some of the BRICS economies are gradually adopting Capitalist strategies for strategic growth akin to those adopted by the West (Murphy et al., 2012). While Brazil collaborates with the 'ABCDs', Asian capital has been invested in companies such as Noble, Olam, Wilmar and Sinar Mas (the 'NOWS') to establish a counter-measure to the market power of Western agribusiness (Keulertz, 2012). One must stress here that the strategic counter-balancing to the ABCDs is particularly driven by Chinese capital to form Asian nodes of power in global agriculture that decrease dependency on Western agribusiness and thus use newly acquired market power to influence global agricultural trade. The restructuring of global food markets is therefore a field of highly strategic ambitions that reflect not only future climate change vulnerabilities but also population, economic and political development. The competition and uncertainty around these international competitive strategies is another source of (geopolitical and market) threat to agriculture.

Population growth

We used CIESIN (2002) SRES B2 gridded population growth scenario for 1990 and 2025 as an estimate of regional population growth trends. The B2 scenario produces a global population of 8.039 billion by 2025. Results indicate significant differences in recent population and in expected growth rate between countries with significant increases on already high populations in India and China, significant increases on currently lower populations in Brazil and South Africa and decreases on an already low population expected for Russia (Table 8.8).

Land degradation

Even in the absence of climate change, land degradation can be a significant impact on agricultural potential. Land degradation can be estimated by examining climate-corrected negative trends in vegetation vigour measured by >LO8 remote sensing using

Table 8.8. Population projections for the BRICS tiles (values in brackets are those for national territory of BRIC country within the tile area where this is different to the whole tile value, i.e., where the tile includes more than one country).

Country	Total population 2007 (Landscan)	Projected percent change (1990–2025)
Brazil	189,276,780	43%
Russia	141,193,660	–11%
India	1,117,329,900 (390,000,000)	62% (58%)
China	1,340,098,400	28%
South Africa	44,023,872	98% (97%)

long term time series of normalised difference vegetation index (NDVI). We used the Bai et al. (2008) analysis of negative trends in RUE-adjusted NDVI (1981–2003) as a surrogate for land degradation and examined: (a) the mean negative trend for areas with a negative trend and (b) the percentage of the tile with a negative trend, as indicators of the degree of recent land degradation. For the Brazil tile, 2% of the area shows degradation with a mean trend of –0.015. For Russia 4% of the area shows degradation with a mean trend of –0.0082. For India 6% of the area shows degradation with a mean trend of –0.013. For China 40% of the area shows degradation with a mean trend of –0.015 and for South Africa 6% of the area shows degradation with a mean trend of –0.013. Clearly China has extensive areas of recent vegetation degradation that is independent of climate variability over the period and potentially very significant for the country's agricultural ambitions. Brazil has very few areas in this situation.

Conclusions

The BRICS (and especially Brazil and India) are huge exporters to many countries and this creates a global interconnectedness and dependency on the agricultural productivity within the BRICS. Changes to water resources for agriculture in these countries will have national impacts, especially for the more populous ones, but also international impacts along the supply chains. Net agricultural virtual water flows out of Brazil and India and into Russia and China, with South Africa in approximate balance. Brazil and India are thus of major concern with respect to food-water security now and in the face of climate change. China and India are also of concern because of the combined impacts of population growth and land degradation (see Mulligan et al., 2011). In addition to their exports of food and food-water, some of the BRICS are also significant exporters of financial investment (see Allan et al., 2012), and of knowledge, technology and skills: South African farmers are increasingly moving up north into other parts of SSA; Brazil as the global champion of revolutionary agricultural expertise. This geographical diversification and cooperation to supply food to BRICS and global markets is a sensible response to climate change and land degradation.

We show that for an ensemble mean Brazil, China and South Africa are likely to see significant negative impacts of climate change whereas Russia and India may see positive impacts. These impacts largely accrue from changes in precipitation with snowmelt and other factors having very small impacts at the regional and national scale. However, because GCM rainfall uncertainty is huge, different statistics of the GCM ensemble (or different GCMs) can show changes in direction (all BRICS drying versus all BRICS wetting) that render GCM climate projections of little use in specific adaptation planning.

While Brazil has a large potential to increase its water exports through utilising current virgin land across the country (though with important environmental consequences), China and India may soon have reached their water export ceiling. Climate change may serve as an aggravating factor for some BRICS, yet Brazil and Russia (with land) and South Africa (with technology) may be able to fill the gap created by future climate change impacts in the already water-stressed and degrading China and India. Under current conditions, Russia has the largest potential of all BRICS economies to increase its agricultural output and thus virtual food-water supply to the global economy. Brazil has much land but its forests are prized for their natural capital and ecosystem services, making them less available for massive agricultural development. South Africa on the other hand may aide Sub-Saharan Africa to better utilise its agricultural potential, thus providing additional virtual water to the global economy. However, development in both Russia and South Africa require the mobilisation of capital into the agricultural sector. If we take both these BRICS economies as crucial for future global water security, investment in farmers is critical. Trade of food and the water embedded in food will not only require investment in farming technology but also improved infrastructure such as railway, roads and storage to allow both economies to grow economically to fill the virtual water gap across the world that may be left by climate change, population growth and land degradation.

This level of uncertainty between GCMs for the same scenario, coupled with uncertainty over emissions scenarios and their controlling factors (population growth, energy prices, economic factors) and agricultural change including land degradation, means that the impacts of climate change on agricultural futures is very difficult to project. Moreover we have examined mean annual and monthly changes whereas agriculture and agricultural productivity is also very prone to changes in the frequency and magnitude of extreme weather events (rainfall, frosts, drought) which are even less predictable. How can we thus manage agriculture in the BRICS for climate-robustness where the future could go one way or the other? For sure we cannot adapt to a specific climate change because we have very little idea of how the climate will change. We therefore have to make agriculture more adaptable and robust to a wide range of changes. This means investments in traditional forms of water harvesting and storage, adaptation of agriculture to local conditions, eco-efficient approaches to minimise the cost and waste of agricultural inputs, including energy and crop diversification and adaptation to local climate and soil conditions. Massive high input monocultures will not help us to adapt to complex climate changes in heterogeneous landscapes. Whilst Brazil, Russia and China have the skills, water and land to buffer some climate impacts in other regions, the effective and sustainable operation of international markets and international co-operation to value water in the food supply chain will be necessary to achieve this.

Acknowledgements

This chapter draws on analysis with the WaterWorld Policy Support System that has been developed over many years under a wide range of EU, CGIAR Challenge Programme on Water and Food and other funding sources. The many providers of global datasets used in WaterWorld and of the COMTRADE database are also gratefully acknowledged.

References

Allan, John Anthony, Keulertz, Martin, Sojamo, Suvi and Warner, Jeroen. 2012. Handbook on Land and Water Grabs in Africa: Foreign Direct Investment and Food and Water Security. Routledge, New York.

Bai, Z.G., Dent, D.L., Olsson, L. and Schaepman, M.E. 2008. Global Assessment of Land Degradation and Improvement 1. Identification by remote sensing. GLADA Report 5. Version 5.

Bokusheva, R. and Hockmann, H. 2006. Production risk and technical inefficiency in Russian agriculture. Eur. Rev. Agric. Econ. (March 2006) 33(1): 93–118.

Center for International Earth Science Information Network (CIESIN). 2002. Country-level Population and Downscaled Projections based on the B2 Scenario 1990–2100 [digital version]. Palisades NY: CIESIN Columbia University. (Available at: http://www.ciesin.columbia.edu/datasets/downscaled.).

FAO. 2011. Climate change, water and food security. FAO Water Reports 36, Rome, FAO (Available at: http://www.fao.org/nr/water/jsp/publications/search.htm).

Friedmann, H. 1987. International regimes of food and agriculture since 1870. pp. 258–276. *In*: Shanin, T. (ed.). Peasants and Peasant Societies. Basil Blackwell, Oxford.

Friedmann, H. 1993. The political economy of food: a global crisis. New Left Review 197: 29–57.

GOI. 2013. State of Indian Agriculture 2012–2013, Department of Agriculture and Cooperation, Ministry of Agriculture, GOI (Government of India), New Delhi (Available at: http://agricoop.nic.in/Annual%20 report2012-13/ARE2012-13.pdf).

Greyling, J.C. 2012. The Role of the Agricultural Sector in the South African Economy. University of Stellenbosch MSc thesis. http://www.agbiz.co.za/LinkClick.aspx?fileticket=k7uLEzO35Fg%3D&tabid=433.

Hijmans, R.J., Cameron, S.E., Parra, J.L., Jones, P.G. and Jarvis, A. 2005. Very high resolution interpolated climate surfaces for global land areas. Int. J. Climatol. 25: 1965–1978.

Keulertz, M. 2012. The Middle Eastern Food Security Question and the Global Food Regime. Conference Paper for Food Security in the Dry Lands, Doha 14-15 November 2012. http://www.fsdl.qa/Portals/0/pdf/16_MARTIN.pdf.

Mekonnen, M.M. and Hoekstra, A.Y. 2011. National water footprint accounts: the green, blue and grey water footprint of production and consumption, Value of Water Research Report Series No. 50, UNESCO-IHE.

Monfreda, Chad, Navin Ramankutty and Jonathan A. Foley. 2008. Farming the planet: 2. Geographic distribution of crop areas, yields, physiological types, and net primary production in the year 2000. Global Biogeochemical Cycles, Vol. 22, GB1022, doi:10.1029/2007GB002947.

Mulligan, M. and Burke, S.M. 2005. FIESTA: Fog Interception for the Enhancement of Streamflow in Tropical Areas. Report to UK DfID, 174 pp. http://www.ambiotek.com/fiesta.

Mulligan, Mark, Fisher, Myles, Sharma, Bharat, Xu, Z.X., Ringler, Claudia, Mahé, Gil, Jarvis, Andy, Ramírez, Julian, Clanet, Jean-Charles, Ogilvie, Andrew and Ahmad, Mobin-ud-Din. 2011. The nature and impact of climate change in the Challenge Program on Water and Food (CPWF) basins Water International, 1941-1707, Volume 36, Issue 1, 2011, Pages 96–124 http://dx.doi.org/10.1080/02508060.2011.543408.

Mulligan, M. 2013. WaterWorld: a self-parameterising, physically-based model for application in data-poor but problem-rich environments globally. Hydrol. Res. 44(5): 748–769. doi:10.2166/nh.2012.217.

Murphy, S., Burch, D. and Clapp, J. 2012. Cereal Secrets: The World's Largest Grain Traders and Global Agriculture. Oxfam Research Reports, Oxford.

Pearce, F. 2013. South Africa's farmers are moving further north, The Guardian, 1 May 2011 (Available at: http://www.guardian.co.uk/environment/2011/may/01/boers-moving-north-african-governments).

Ramankutty, N., Evan, A.T., Monfreda, C. and Foley, J.A. 2008. Farming the planet: 1. Geographic distribution of global agricultural lands in the year 2000. Global Biogeochemical Cycles Vol. 22 GB1003 doi:10.1029/2007GB002952.

Shah, T., Singh, O.P. and Mukherji, A. 2006. Some aspects of South Asia's groundwater irrigation economy: analyses from a survey in India, Pakistan, Nepal Terai and Bangladesh. Hydrological Journal 14: 286–309.

Siebert, S., Burke, J., Faures, J.M., Frenken, K., Hoogeveen, J., Doll, P. and Portmann, F.T. 2010. Ground water use for irrigation—a global inventory. Hydrol. Earth. Syst. Sci. 14: 1863–1880.

Sojamo, S., Keulertz, M., Warner, J. and Allan, J.A. 2012. Virtual water hegemony: the role of agribusiness in global water governance. Water International 37(2): 169–182.

Tebaldi, C. and Knutti, R. 2010. Climate models and their projections of future changes. pp. 31–56. *In*: Lobell, David B. and Burke, Marshall (eds.). Climate Change and Food Security: Adapting Agriculture to a Warmer World. Springer, Amsterdam.

Teubal, M. 1993. Agroindustrial modernization and globalization: towards a new world food regime. Institute of Social Studies, Working Paper Series No. 162.

United Nations Commodity Trade Statistics Database, Department of Economic and Social Affairs/ Statistics Division (2013)n COMTRADE database. http://comtrade.un.org/db/.

United States Department of Agriculture Foreign Agricultural Service. 2013. Russia: Agricultural Overview. USDA: Washington D.C. (Available at: http://www.fas.usda.gov/pecad2/highlights/2005/03/Russia_Ag/).

Walsh, P.D. and Lawler, D.M. 1981. Rainfall seasonality: description, spatial patterns and change through time. Weather 36: 201–208.

Warner, J., Sebastien, A. and Empinotti, V. 2013. Claiming back the land: the geopolitics of Egyptian and South African land and water grabs. pp. 223–243. *In*: Allan Tony, Keulertz Martin, Sojamo Suvi and Warner Jeroen (eds.). Handbook of Land and Water Grabs: Foreign Direct Investment and Food and Water Security. Routledge, London.

World Bank. 2013. India: Issues and Priorities for Agriculture (Available at: http://www.worldbank.org/en/news/feature/2012/05/17/india-agriculture-issues-priorities).

9

Trading Off Agriculture with Nature's Other Benefits, Spatially

Mark Mulligan

Introduction

We use the Co$ting Nature policy support system [v2.48] (Mulligan, 2013; Mulligan et al., 2010) with 1 km spatial resolution input datasets from various sources focused around the year 2000, to calculate the realized ecosystem service provision based on current (2010) land cover and use. We examine the spatial correlation between the magnitude of different services and the current distribution of cropland, pasture and their productivities.

We then develop a scenario for land cover and use change on the basis of continued agriculturalisation in these areas at recently observed rates. We analyse the trade-offs between gains in agricultural productivity and losses in nature's other benefits under this scenario for the two very different biophysical and socio-economic contexts of Colombia and MatoGrosso.

The Co$ting Nature Policy Support System

Co$ting Nature assumes that conservation and development priority should be defined on the basis of data for ecosystem services, biodiversity, current pressure and future threat. Clearly, the development value of land is also important to such decisions, especially suitability for agriculture and potential productivity as well as the potential value of the mineral, forestry and other natural resources. These values need to be

Department of Geography, King's College London, Strand, LONDON, WC2R 2LS.
Email: mark.mulligan@kcl.ac.uk

seen within the context of the costs of developing and maintaining their production or extraction but also relative to conservation priority: for example the trade-off between protection of a high conservation value area and its value for rare-earth mineral mining may be different to the tradeoff between high conservation value and low-productivity agricultural development potential.

Biodiversity and ecosystem services metrics can be estimated in biophysical units, but these units cannot easily be combined to assess the bundle of services at a site. Some authors combine services by converting them to a common economic currency through valuation (Costanza et al., 1997; Troy and Wilson, 2006), others present them separately (Raudsepp-Hearne et al., 2010). In Co$ting Nature we combine services by expressing each as a normalised value between the globally or the locally highest and lowest values. The normalised values (expressed in values of 0–1) can then be combined irrespective of their units. The analyst can define whether values are normalised to a global maximum as would be necessary for global comparisons or a local maximum to more clearly highlight local patterns. For this paper we normalise locally to represent local patterns most clearly.

Carbon

Our carbon storage and sequestration service is calculated as the combination of carbon stocks and carbon sequestration. Carbon sequestration is calculated here from the dry matter productivity (DMP) analysis of Mulligan (2009b) in which SPOT-VGT DMP calculated every 10 days at 1 km resolution on the basis of change in NDVI, was averaged over the period 1998–2008, globally. DMP (t biomass/ha/yr) is multiplied by 0.42 (Ho, 1976) to convert to units of tC/ha/yr. Above-ground carbon stock is calculated from Saatchi et al. (2011) for the areas in which data are available and Ruesch and Gibbs (2004) elsewhere. This is combined with soil carbon calculated from the map of Scharlemann et al. (2009) to produce total above- and below-ground carbon stocks.

Water

Water is considered a provisioning service in Co$ting Nature, though it also plays a role in the regulating services (see section on hazard mitigation). Potential water services are measured as the volume of runoff (rainfall minus evapotranspiration) whose quality is unaffected by human activity, cumulated downstream. This is an indicator of the volume of clean water produced by a pixel. The human footprint on water quality index (HF) (Mulligan, 2009) is used as the indicator of water quality. The HF considers particular land uses to have the potential to contaminate water with sediment, agrochemicals, manures, etc. Land uses such as unprotected agriculture and pasture are considered non-point sources and roads, mines, oil and gas wells and urban areas are considered point sources. Agriculture in protected areas is considered to have a human footprint of zero as are areas with no human land use. The HF index multiplies the water balance of a pixel by the fractional cover of point and non-point sources in that pixel and cumulates this 'polluted' water downstream using a streamflow network. The total volume of water flowing is also cumulated downstream and the

HF in a pixel becomes the volume of polluted water as a percentage of the total water flowing. The non polluted volume of water expressed as a fraction is considered the potential water service of a pixel. Realized water services are calculated considering the use of these potential water services which, in turn, depends on the distribution of beneficiaries for hydrological services.

Maps of globally normalised downstream population, irrigation area and number of dams are pre-calculated to represent the distribution of beneficiaries. Population is derived from Landscan (2007) and is summed downstream at 1 km resolution using a drainage network derived from HydroSHEDS (Lehner et al., 2008) with the downstream total for a pixel assigned to that pixel. For number of dams we use the global dam census of Mulligan et al. (2011) and for irrigated land we cumulate the downstream irrigated areas of Siebert et al. (2007) for each point. These provide pixel-level indicators of the distribution of the beneficiaries of hydrological ecosystem services at a much finer resolution than previous global studies. Potential water services are multiplied by the normalised sum of all downstream beneficiaries to calculate the realized water services index.

Hazard mitigation

Hazard mitigation services are perhaps the most complex to assess since they are a function of:

a) the potential for multiple hazards to occur and the human and infrastructural exposure to those hazards and vulnerability to the negative impacts of hazards. Exposure and vulnerability together define the risk.
b) the role of local, upstream or near-coast ecosystems in reducing the potential impact of hazards (i.e., potential hazard mitigation services).

Hazard mitigation ecosystem services are then realized in those areas in which ecosystems provide hazard mitigation services but where there is also risk. Areas with no risk may receive potential hazard mitigation services but these services are not realized. Co$ting Nature considers hazard potential as:

a) the normalised frequency of cyclones according to Dilley et al. (2005) multiplied by the normalised water balance as an index of high magnitude rainfall event hazards.
b) for coastal inundation hazards we calculate distance from coast according to USGS (2006) and consider all pixels within 2 km as coastal. We also produce an index of lowlying land as all areas from 0 to 30 m according to the SRTM digital elevation model post-processed by Lehner et al. (2008). The probability of (coastal) inundation hazard is considered proportional to the normalised probability of Tsunami (according to NGDC, 2011), cyclones, and climatic sea level rise (considered for simplicity as equally likely everywhere) for all coastal areas. The probability of inundation hazard is the combination of these effects.
c) for landslide hazards we consider the probability of landslides to increase with the normalised mean upstream slope gradient. Upstream slope gradient is pre-

calculated using the 1km resolution digital elevation model and flow network of Lehner et al. (2008).

d) the potential for flooding is considered proportional to normalised water balance with small potential in dry areas and high potential where water is plentiful. Though many floods are fluvial in nature, we use water balance rather than runoff in recognition that floods also occur from overwhelmed urban drainage, groundwater flooding and rainfall intensity greater than infiltration rates. These latter types of flood can be somewhat more predictable than fluvial floods.

Hazard potential is thus the mean of cyclone, inundation, landslide and flood probabilities, normalised either locally or globally as defined by the user. In addition to hazard potential, we consider hazard exposure as the exposure of human populations, activity and infrastructure. Socio-economic exposure is calculated as normalised GDP for 1990 (CIESIN, 2002), population (Landscan, 2007), agriculture (cropland and pasture fractional areas from Ramankutty, 2008), and infrastructure. Infrastructural exposure is calculated as the sum of the presence of dams (Mulligan et al., 2011), mines (Mulligan, 2010a), oil and gas (Mulligan, 2010b), urban areas (Schneider et al., 2009) and roads (FAO, 2010). Exposure is multiplied by hazard potential to produce the index of exposure to hazards. Vulnerability to hazards is considered to scale with normalised GDP and infrastructure: the greater the GDP and infrastructure, the greater the capacity to cope with hazards. Risk is then exposure multiplied by vulnerability.

Potential hazard mitigation *services* are then calculated according to a series of assumptions, based on knowledge of how ecosystems mitigate these hazards. We assume that landslide impacts at a point are mitigated according to the proportion of upstream area that is tree covered (using tree cover data from Hansen et al., 2006) or is protected (according to WDPA, 2012). This, because tree cover reduces potential soil waterlogging and has been shown to reduce landslide frequency (Dapples et al., 2002) and protected areas will tend to have a lower agricultural and infrastructural impact—both of which can lead to increased frequency of landslides. Regulation of drought hazards (for example reduced dry season flows) are assumed proportional to tree cover upstream. Although trees evaporate significant volumes of water and thereby reduce flows, they also encourage infiltration that helps to maintain dry season flows (Pena Arancibia et al., 2012). Flood hazards are mitigated according to the proportion of upstream area that provides flood storage in the form of trees, wetlands (Lehner and Doll, 2004), water bodies (USGS, 2006) and floodplains (Mulligan, 2010). Mitigation from coastal inundation is considered to be provided by wetlands and mangroves (Spalding et al., 1997) but only in low-lying and coastal areas. Where they occur inland they are assumed to have no coastal inundation mitigation potential. The total potential hazard mitigation services is then the mean of coastal, flood regulation and landslide mitigation services.

Realized hazard mitigation services are calculated as the minimum of risk and potential hazard mitigation services for areas where risk is greater than 0. In other words if hazard mitigation potential is greater than risk, then hazard mitigation potential equals risk and the remaining hazard mitigation potential is unused. If risk is greater than hazard mitigation potential then realised hazard mitigation is equal to potential hazard mitigation and some risk remains unmitigated.

Biodiversity

Focusing on ecosystem services alone ignores the reality that biodiversity is an important factor determining nature conservation priority since it has intrinsic value (Oksanen, 1997; Ghilarov, 2000; Justus et al., 2009). Biodiversity also supports a range of ecosystem services including pest control (Bianchi et al., 2006), ecosystem stability (Tilman et al., 2006) and plant genetic resources. The most conservation-relevant measure of biodiversity is difficult to define: ecosystem diversity, genetic diversity, taxonomic diversity and evolutionary distinctiveness are all candidates but given data constraints most studies use a measure of species richness (Purvis and Hector, 2000). In Co$ting Nature we combine a measure of species richness with a measure of species endemism. Endemism captures rarity, which is clearly of conservation relevance in addition to species richness.

Species richness

We use the IUCN sampled redlist extent of occupancy (EOO) data for amphibians (IUCN, 2008), mammals (IUCN, 2008b), reptiles (IUCN, 2010) and birds (Birdlife, 2012) and calculate a measure of richness (total number of sampled species within each 10 km) combining all sampled species. We would have liked to incorporate similar data for plants but these data are not in the public domain.

Endemism richness

Kier and Barthlott (2001) define endemism richness as the "specific contribution of an area to global biodiversity". Since it accounts for range size rarity, it is potentially a better measure of conservation value than species richness. The contribution of a specific area to the global species inventory is known as its C-value. For the 10 by 10 km (native) raster grids used here the C-value of a pixel for a given species was calculated as 1/G, where G is the global range size (i.e., the number of pixels in which the species occurs). Thus where species have a large range, the C-value is low, and where their range is restricted, the C-value is high. The endemism richness of a site is thus the sum of C-values for all the taxa for which data are available. The species and endemism richness metrics are both normalised and then combined into a single normalised biodiversity metric for display.

Current Agriculture, Agricultural Suitability and Ecosystem Service Provision in Brazil and Colombia

Having described Co$ting Nature, we now apply the tool to analyse ecosystem service provision in Brazil and Colombia in relation to agriculture and agricultural suitability.

Colombia

Figure 9.1 shows the spatial relationship between a range of ecosystem services and current cropland productivity and suitability. Carbon storage (a) (above and below ground) and sequestration are highest in areas of highest soil carbon (peat soils) in the Andes but also in areas of high above-ground storage and sequestration of carbon in the Amazon forests. Hydrological ecosystem services (b) are highest in the high rainfall and undisturbed areas upstream of the greatest populations, dams and areas of irrigation, i.e., the headwaters of the Magdalena Valley. Hazard mitigation services (c) are highest in the areas with significant cropland, pasture and infrastructure that generates exposure to the flood, landslide and soil erosion hazard potential that exists in these areas. Biodiversity (d) is highest in the Amazon lowlands, the Pacific lowlands and the inter-Andean valleys. These patterns show some complementarity but also quite some differences in their geographical distributions.

Figure 9.1 also shows that intensive cropland (e) is generally concentrated in the inter-Andean valleys and especially in the central and south regions. Less intensive cropland (e) is also to be found throughout the Atlantic lowlands and in the Llanos savannas of the north east. Pastures (f) are much more widespread on the Andean slopes, throughout the Atlantic lowlands and in the Llanos. Crop suitability (48 crops) based on FAO (2012), is highest in the inter-Andean valleys and the Llanos for both high (g) and low (h) input agriculture. There is certainly significant overlap between current agriculture and areas with high ecosystem service provision, though current agriculture is largely not in the areas of greatest biological diversity. Crop productivity

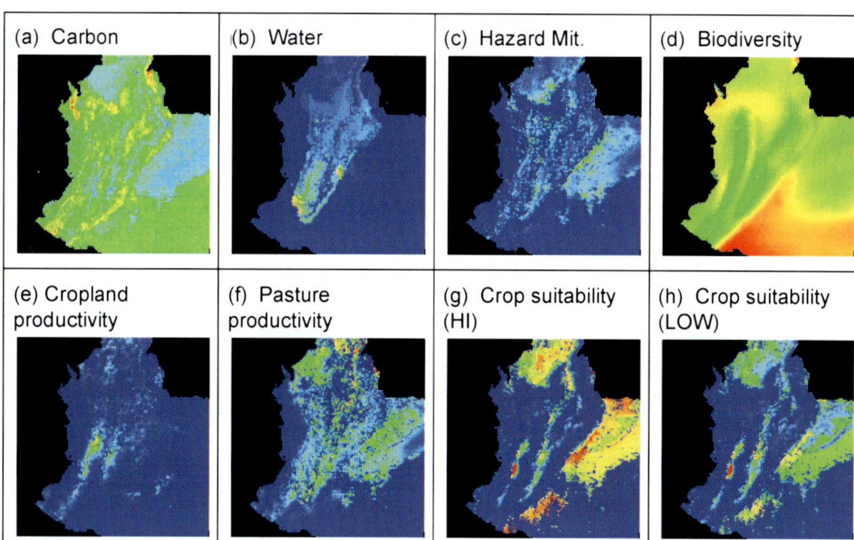

Figure 9.1. Distribution of ecosystem services and crop productivity in Colombia.

is calculated based on the productivity (DMP) analysis of Mulligan (2009b) multiplied for each 1 km pixel by the fractional cover of cropland or pasture from Ramankutty et al. (2008).

Figure 9.2 shows the relationships between individual ecosystem services (columns), dry matter productivity of croplands (first row), dry matter productivity of pastures (second row), crop suitability (high inputs, third row) and crop suitability (low inputs, fourth row). There are clearly few strong relationships between individual services and crop agricultural productivity and suitability at a pixel scale.

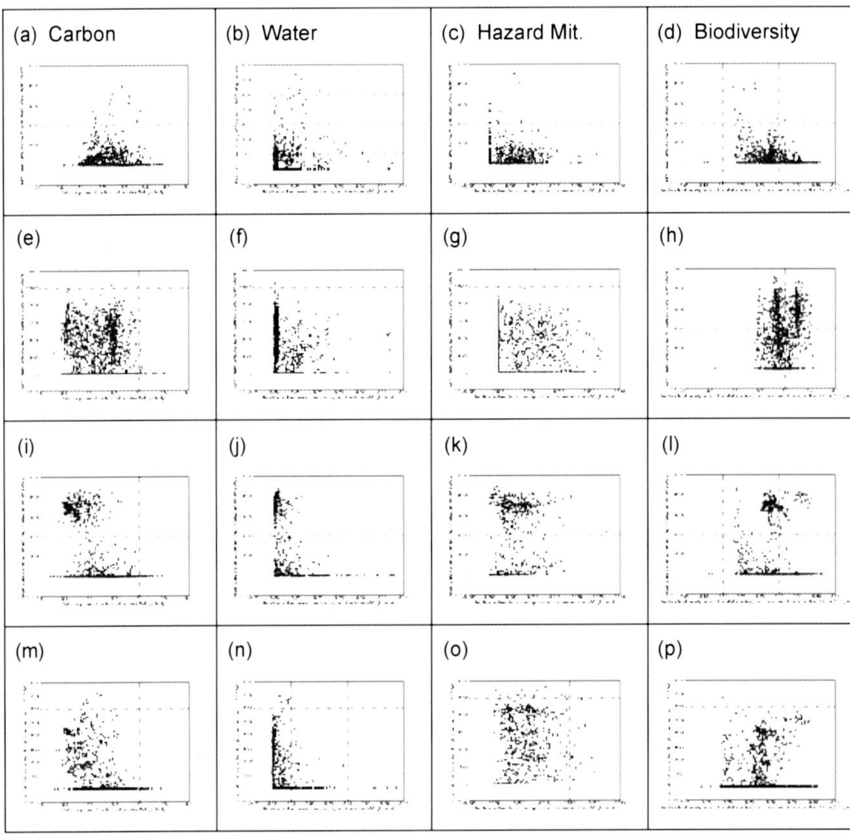

Figure 9.2. Relationships between crop suitability, productivity (X-axis) and provision of non-agricultural ecosystem services for Colombia (Y-axis).

Brazil

For MatoGrosso, Brazil (Fig. 9.3), carbon storage and sequestration (a) is highest in the north-west where rainforests remain intact, whereas realized hydrological services (b) are highest towards the south which are more populous, have more agriculture and in which, therefore, more of the potential services are realized. Hazard mitigation services (c) follow a similar pattern (high to the south and east) in which the realized service

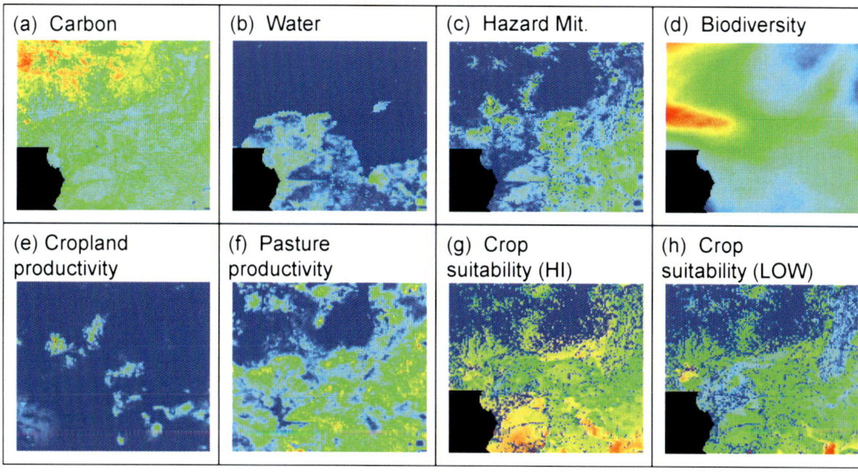

Figure 9.3. Distribution of ecosystem services and crop productivity in Colombia.

is much greater in the more agricultural and populous areas that are hazard-prone which mean that potential hazard mitigation services are then realized. Biodiversity is highest in the north-west and Andean foothill forests rather than the southern savannas.

Cropland (e) is intensive in a few areas while pasturelands (f) are extensive in the south and east. Crop suitability (g and h) is high throughout the south whether under high or low inputs and thus there is great potential for further expansion, though with risks to water and hazard mitigation services given their geographical overlap.

Figure 9.4 shows relationships between individual ecosystem services (columns) and dry matter productivity of croplands (first row), dry matter productivity of pastures (second row), crop suitability (high inputs, third row) and crop suitability (low inputs, fourth row) for MatoGrosso. Though the majority of ecosystem services have weak relationships with agricultural productivity or suitability on a pixel level in MatoGrosso, there are some positive relationships, for example higher water services are associated with lower dry matter productivity of pastures, whereas higher hazard mitigation services show the opposite relationship (increasing with pasture productivity). High pasture productivity in these intensive pastures reduces water services because of the impact of pastures on water quality (through Co$ting Nature's human footprint index). Hazard mitigation services increase with pasture productivity because of the greater exposure to hazard potential (and thus realization of available mitigation services) where there are significant agricultural investments.

Agriculture and Ecosystem Service Provision in Brazil and Colombia by 2100

We now carry out a scenario experiment with Co$ting Nature. We develop a land use change scenario using Co$ting Nature's land use change model to project recently observed deforestation rates, according to MODIS (Hansen et al., 2006)

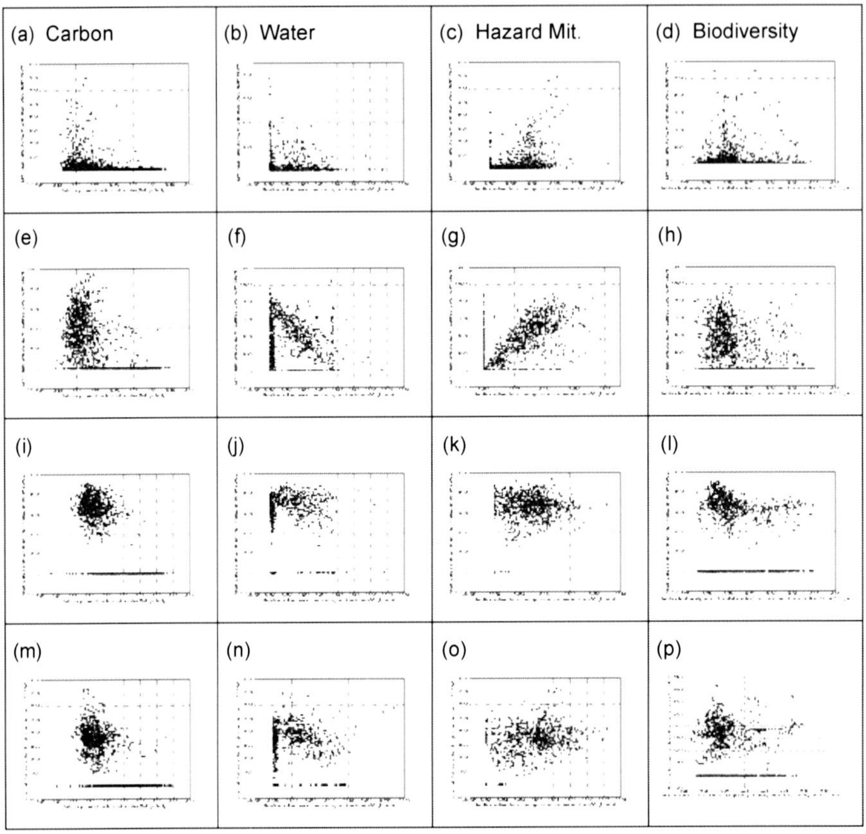

Figure 9.4. Relationships between crop suitability, productivity (X-axis) and provision of non-agricultural ecosystem services (Y-axis) for MatoGrosso, Brazil.

and terra-i (www.terra-i.org) data, forward to 2100. The historic rates are calculated by regional administrative area and the resulting areas deforested within 100 years are allocated according to proximity to existing deforestation fronts, accessibility to planned (IIRSA)[1] and likely transport routes (lines connecting urban areas) as well as agricultural suitability (low inputs). In this scenario, designated protected areas are considered to have no arresting impact on allocation of deforestation. This leads, in Colombia, to large-scale forest loss in the S Amazon, the Pacific and the north-east (Fig. 9.5). Clearly this scenario represents a significant change in land cover and use at the national scale but one which is comparable to those observed historically over much shorter periods of time in nearby countries.

The same scenario in MatoGrosso (Fig. 9.6) ensures that many of the forests present currently in the north-west are replaced by cropland, leaving very few remaining forested patches.

[1] www.iirsa.org

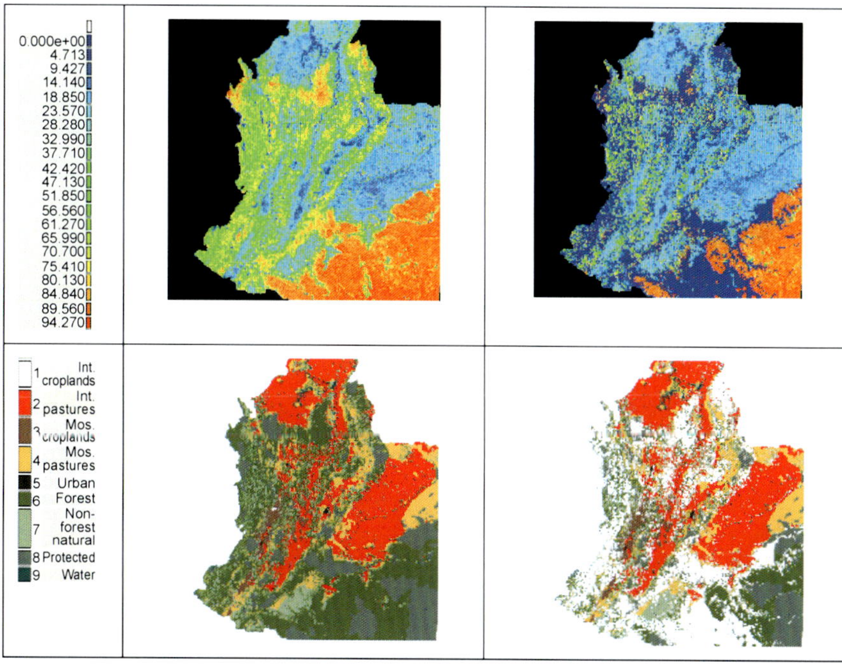

Figure 9.5. Baseline (top-left) and scenario (top-right) tree cover (%) for Colombia and baseline (bottom left) and scenario (bottom right) land use for Colombia.

Though in this scenario deforested areas are converted to intensive cropland land use, the land use could equally have been set to pasture and given the assumed equal human footprint contribution of cropland and pasture land uses in Co$ting Nature and the similar change in land cover through removal of trees and their replacement with herbaceous and bare cover, the following results would apply broadly whether the land use applied were cropland or pasture.

Co$ting Nature calculates the impact of such land use change scenarios on ecosystem services based on the identification of analogous areas. Analogous zones for land cover and use change are identified as follows: mean annual temperature and mean annual precipitation are each zoned into 10 equal interval classes spanning their range which, when combined, can produce as many as 100 class combinations. Within each of these temperature and precipitation zones we identify pixels that have in the baseline the tree and herb cover values (within 5%) that are assigned in the scenario. We then transfer baseline ecosystem service values (for those services affected by land use change) from the baseline zones to the equivalent zones in the scenario. In this way the value of, say, carbon storage in an area converted from tree cover to herb cover is assigned by identifying the mean value of carbon for all herbaceous cover pixels in the same climate zone (according to temperature and precipitation). This value is then assigned to all pixels in the same climate zone that have changed to that herbaceous

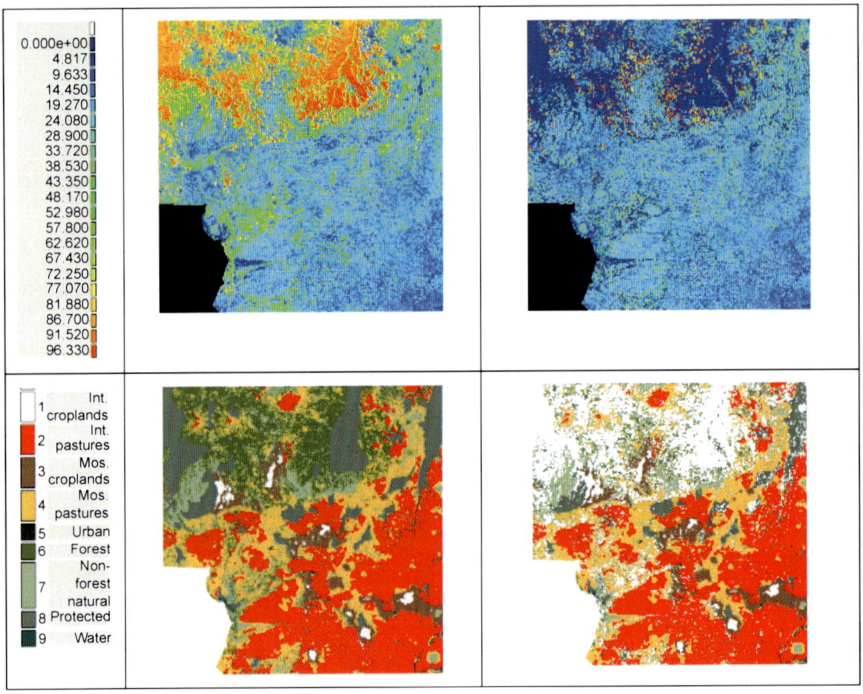

Figure 9.6. Baseline (top-left) and scenario (top-right) tree cover (%) for MatoGrosso and baseline (bottom left) and scenario (bottom right) land use for MatoGrosso.

cover fraction. This simple value transfer method is the only feasible approach given the complexity and variety of ecosystem services modelled by Co$ting Nature.

Trade-offs between agricultural suitability and ecosystem service provision

Here we examine the impact of this land use change scenario on ecosystem service provision and agriculture in the two study areas. In both cases cropland cover increases by around 300 K square-km affecting directly 2.6 M and 4.1 M people currently living in the areas to be converted to intensive cropland in Colombia and Brazil respectively. Carbon storage services in Colombia is reduced by around 3% and in Brazil by around 8.3% (higher because of a low baseline storage in Brazil). Carbon sequestration similarly declines. Water services are much more significantly affected since these are concentrated in the human-occupied areas that are closest to the areas of land use change (whereas many of the carbon services provided are by areas remote from people). Realized water services decline by 32% in Colombia affecting some 26 M people locally and downstream. They decline by 25% in Brazil affecting 540 K people in the much less densely populated context of MatoGrosso.

Table 9.1. Change in ecosystem services based on a scenario of continued agriculturalisation with no conservation of protected areas in place.

Variable	Colombia	Brazil
Change in cropland cover	290 K(2.6 M)	330 K(4.1 M)
Change in pasture cover	0	0
Change in carbon storage	–3%	–8.30%
Change in carbon sequestration	–3%	–4.20%
Change in water services index	–32% (26 M)	–25% (540 K)
Change in hazard mitigation index	+8.5% (580 K)	+1.6% (4700)
Change in biodiversity index	–0.13 (680 K)	–0.21 (58 K)

Changes to ecosystem services: Colombia

In Fig. 9.7 (a) to (e) below, we examine the spatial distribution of change in cropland and ecosystem services with this scenario. Ecosystem services range from blue (low) to red (high) in all maps. Clearly the distribution of cropland increases significantly in the Amazon and Pacific but also in parts of the Andes.

This leads to reduction in carbon services concentrated in the Andean hills and foothills and in the Caribbean plains.

Hydrological ecosystem services are also affected in the areas converted to croplands (because of the resulting decrease in water quality and also potential losses of flow in areas where cloud forests are lost, see Bruijnzeel et al., 2011), the reductions are particularly clear in the Andes and especially the upper parts of the Magdalena river valley.

Hazard mitigation services increase throughout the newly agriculturalised areas, largely as a result of the increasing socio-economic and infrastructural exposure to hazards that results from the agricultural investments. The increases in hazard mitigation services are greatest in the areas with greatest hazard potential, for example parts of the Pacific that are exposed to very high rainfall or coastal hazards, the steep eastern Andes and the flood-prone lower Magdalena.

Biodiversity is also affected by the conversion to cropland: not all biodiversity in the cultivated areas is lost but there is a significant change, especially in the Amazon.

Changes to ecosystem services: Brazil

In Fig. 9.8 (a) to (e) below we examine the spatial distribution of change in ecosystem services with this same scenario for MatoGrosso, Brazil. Ecosystem services range from blue (low) to red (high). Clearly, the distribution of cropland increases significantly in the north and west of MatoGrosso where there is still capacity for expansion.

This leads to significant decreases in carbon storage and sequestration in this currently forested area, though significant carbon storage and sequestration remains elsewhere (in soils, remaining vegetation cover and sequestration—even by agriculture).

Though much of the land use change occurs in the north-west of the region, the realized hydrological ecosystem services change little here because of the low

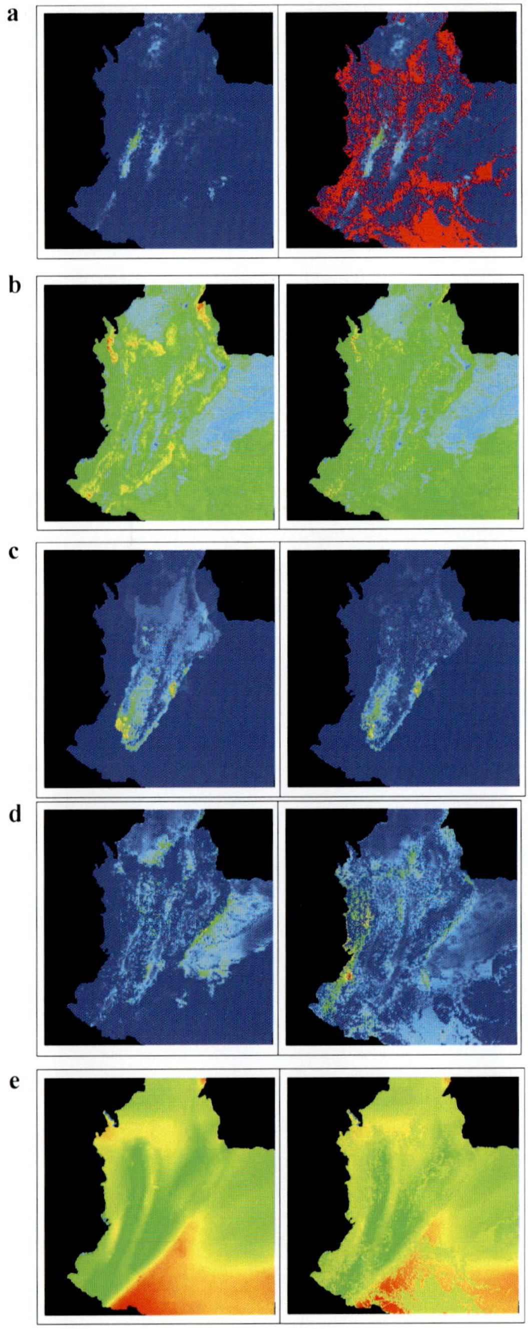

Figure 9.7. (a) Cropland baseline (left) and scenario (right). **(b)** Carbon baseline (left) and scenario (right). **(c)** Water baseline (left) and scenario (right). **(d)** Hazard Mitigation baseline (left) and scenario (right). **(e)** Biodiversity baseline (left) and scenario (right).

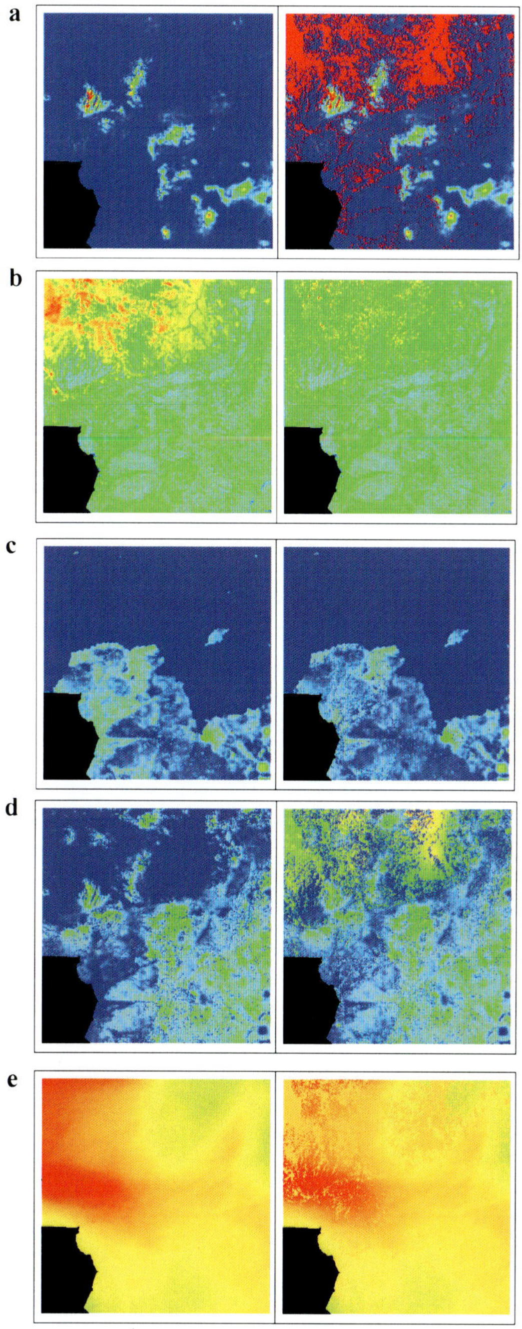

Figure 9.8. **(a)** Cropland baseline (left) and scenario (right). **(b)** Carbon baseline (left) and scenario (right). **(c)** Water baseline (left) and scenario (right). **(d)** Hazard Mitigation baseline (left) and scenario (right). **(e)** Biodiversity baseline (left) and scenario (right).

populations and other uses of water downstream of this region compared with further south. Potential hydrological services decrease significantly but since these services are not realized they are not missed. Clearly if population were to increase significantly in the region or downstream in line with the cropland expansion then realized services would also decrease significantly here. Decreases are much more obvious further south where realized services are greater to start with.

Hazard mitigation services increase significantly in the areas where new cropland appears, because of the increase in socio-economic exposure to the extant hazards resulting from the agriculturalisation of these areas.

Biodiversity also decreases where formerly forested landscapes are converted to cropland, especially in the high biodiversity regions of the north and west.

Clearly there are complex geographic patterns of ecosystem service change in relation to the complexities of ecosystem service distribution, of land use change and of the impact of land use change on each service, in relation to the distribution of beneficiaries.

Managing these tradeoffs

If we examine the ratio of these ecosystem services to normalised crop suitability for Colombia (Fig. 9.9), values >1 indicate greater agricultural than ecosystem services ranking. Values <1 (black) indicate greater ecosystem services ranking than agricultural suitability ranking. Clearly over much of the Andes and Amazon the ES value outranks the crop suitability, only in the Llanos, northern plains and the Andean valleys is the opposite true. This helps indicate the *real value* of Andean hillsides, the eastern Andes and the Amazon in the provision of ecosystem services rather than in agriculture on poorly suitable lands.

The ratio of ecosystem services to crop suitability (Fig. 9.10) for MatoGrosso shows areas with values <1 (black) indicating greater ecosystem services than agricultural ranking and blue areas where agricultural value outranks ecosystem

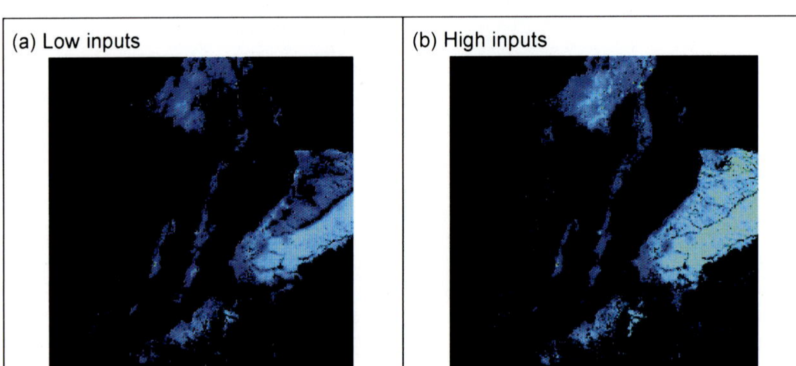

Figure 9.9. Areas where ratio of agricultural suitability to bundled ecosystem services is >1: Colombia for low and high input agriculture.

service value. We have many more areas where the normalised agricultural suitability is greater than that of ecosystem services, especially under high inputs.

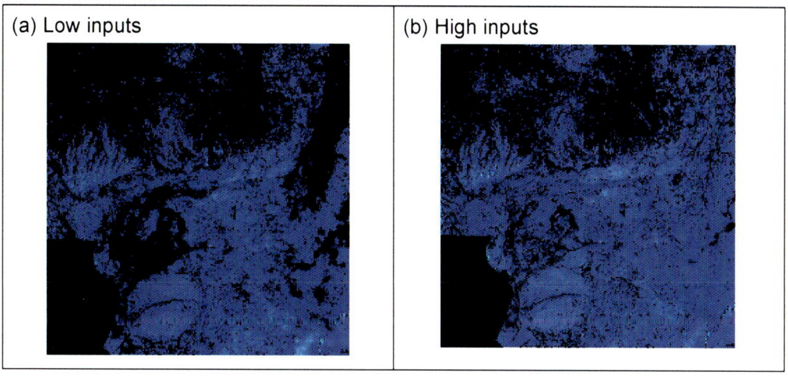

Figure 9.10. Areas where ratio of agricultural suitability to bundled ecosystem services is >1: Brazil for low and high input agriculture.

Implications of Arresting Deforestation in Protected Areas

A further land use change scenario was produced in which agricultural development is not permitted within currently designated protected areas as defined in the WDPA (2012). Since regional rates were held at recent historic levels this will have displaced deforestation outside of protected areas (see Table 9.2). In both Colombia and Brazil

Table 9.2. Comparison of ecosystem services impacts of protected and unprotected scenarios for Colombia and MatoGrosso, Brazil. Figures in square brackets indicate the change per 1000 square km of new cropland.

Variable	Colombia (protected)	Colombia (unprot)	Brazil (protected)	Brazil (unprot.)
Change in cropland cover	260 K	290 K	230 K	330 K
	(2.5 M)	(2.6 M)	(110 K)	(130 K)
Change in pasture cover	0	0	0	0
Change in carbon storage	−2.3% [−0.01]	−3% [−0.10]	−4.9% [−0.02]	−8.3% [−0.03]
Change in carbon sequestration	−1% [0.00]	−3% [−0.10]	−2.8% [−0.01]	−4.2% [−0.01]
Change in water services index	−28% [−0.11]	−32% [−1.10]	−18% [−0.08]	−25% [−0.08]
	(11 M)	(11 M)	(500 K)	(540 K)
Change in hazard mitigation index	0.0043	0.0049	0.0011	0.0017
	+7.4% [0.03]	+8.5% [0.29]	+1.2% [0.01]	+1.6% [0.00]
	(2.5 M)	(2.6 M)	(110 K)	(130 K)
Change in biodiversity index	+0.051 [0.000]	−0.13 [−0.004]	−0.083% [0.00]	−0.21% [−0.001]
	(650 K) –ve	(680 K)	(45 K)	(58 K)

this leads to significant decreases in deforestation compared with the unprotected scenario—even after displacement—because of the reduced availability of suitable land and, so fewer people are affected directly by the conversion to cropland (i.e., are situated in the areas that convert to cropland). The protected scenario leads to an order of magnitude lower declines in carbon storage per square km of new cropland in Colombia under the protected scenario compared with the unprotected one. The carbon storage difference between scenarios is much less per square-km but much more overall for the MatoGrosso case. The same pattern emerges for carbon sequestration. For water, the protected scenario leads to a lower decline in ecosystem services but, per unit of new cropland cover, the loss is greater for the protected than the unprotected scenario (in Colombia at least). This results from the concentration of agriculturalisation in already human impacted landscapes away from protected areas in the protected scenario meaning this change is closer to people who, as a result, are more impacted.

For hazard mitigation, the protected scenario leads to a lower increase in hazard mitigation services than the unprotected scenario with a significant reduction in the gain per square km of cropland in the protected scenario for Colombia. This results from the reduced agriculturalisation and also the focusing of new cropland on areas in which human exposure to hazards already exists. Of course, gains in the use of hazard mitigation services are false gains since all development will lead to such gains.

The protected scenarios lead to a significant reduction in biodiversity index loss for both Colombia and MatoGrosso in comparison with the no protection scenario. The protected scenarios also lead to much lower loss of biodiversity per unit of new cropland than the unprotected scenarios.

Implications of Arresting Deforestation in the Top 17% of Areas by Bundled Ecosystem Services (targeting)

We now apply a further scenario designed to understand whether the current protected areas network achieves a greater protection of non-agricultural ecosystem services compared with a protection network based on **targeting** the top 17% of areas by bundled realised ecosystem services. 17% is the Aichi 2020 biodiversity target for protected areas.[2] The land use change model is run as before but this time denying land use change in these top 17% of areas by bundled ecosystem services.

Compared with the protected scenario, the targeted scenario leads to less cropland development in both Colombia and MatoGrosso because of greater constraint on development in areas with historically high rates of land use change. It achieves a smaller decrease in carbon storage both in total and per unit area of cropland and even a small gain in carbon storage (relative to the baseline) for Brazil, likely an artefact of the benefit transfer approach. Carbon sequestration declines less in total and less per unit of cropland for the targeted scenario compared with the protected one.

Water services also decline less overall in the targeted scenario and less per unit of new cropland for both countries, reflecting the focus of ecosystem service conservation.

[2] http://www.cbd.int/sp/targets.

Table 9.3. Comparison of ecosystem services impacts of protected and targeted scenarios for Colombia and MatoGrosso, Brazil. Figures in square brackets indicate the change per 1000 square km of new cropland.

Variable	Colombia (targeted, low inputs)	Colombia (protected)	Brazil (targeted, low inputs)	Brazil (protected)
Change in cropland cover	220 K	260 K	150 K	230 K
	(1.5 M)	(2.5 M)	(86 K)	(110 K)
Change in pasture cover	0	0	0	0
Change in carbon storage	−1.1% [−0.004]	−2.3% [−0.01]	+0.26% [0.001]	−4.9% [−0.02]
Change in carbon sequestration	−0.19% [−0.001]	−1% [0.00]	−1.2% [−0.005]	−2.8% [−0.01]
Change in water services index	−24% [−0.092]	−28% [−0.11]	−12% [−0.052]	−18% [−0.08]
	(5.2 M)	(11 M)	(490 K)	(500 K)
Change in hazard mitigation index	0.0036	0.0043	0.00067	0.0011
	+4.2% [0.016]	7.40%	+1.1% [0.005]	1.20%
	(1.5 M)	[0.03]	(86 K)	[0.01] (110 K)
		(2.5 M)		
Change in biodiversity index	−0.95% [−0.004]	0.051	+0.16% [0.001]	−0.083% [0.00]
	(520 K)	[0.000]	(36 K)	(45 K)
		(650 K) −ve		

Hazard mitigation services also increase less under the targeted scenario both in total and per unit of cropland.

In the case of Colombia, biodiversity losses are greater in the targeted scenario compared with the protected scenario, both totally and per unit croplands. This reflects the fact that targeting ecosystem services tends to focus conservation on human dominated areas more than the current network of protected areas, which tend to protect more of the world's wilderness areas and thus biodiversity.

Conclusions

The spatial relationships between agriculture, biodiversity and ecosystem services are complex and differ between regions and services. We found very different distributions of carbon, water and hazard mitigation services, with carbon services being greatest away from significant populations but both water and hazard mitigation services being highly dependent on the presence of local human beneficiaries.

We found few relationships between cropland suitability or productivity and ecosystem services in Colombia and MatoGrosso, Brazil. However in MatoGrosso, higher water services are associated with lower dry matter productivity of pastures, whereas higher hazard mitigation services show the opposite relationship (increasing with pasture productivity). High pasture productivity in these intensive pastures reduces water services because of the impact of pastures on water quality (through Co$ting Nature's human footprint index). Hazard mitigation services increase with pasture productivity because of the greater exposure to hazard potential (and thus realization of available mitigation services) where there are significant agricultural investments.

In a deforestation scenario assuming ineffective protected areas policies, cropland cover increased significantly leading to significant decreases for all ecosystem services except hazard mitigation. Hazard mitigation services increase because of the greater realisation of potential services with more agricultural infrastructure. Protecting the current protected areas system into the future led to a reduction in the total conversion to agriculture but also declines in the loss of all ecosystem services and affected populations. These declines were effective in total and per unit area of new cropland. A modified scenario in which the top 17% of land on the basis of baseline bundled ecosystem service provision was protected instead of the current protected areas system led to lower agriculturalisation and a lower loss of ecosystem services in total and per unit area of new cropland. Biodiversity shows a higher decline (in total and per unit of new cropland) in the targeted scenario than the protected for Colombia but the opposite pattern for Brazil. Increased protection of ecosystem services is thus at the expense of the protection of biodiversity in these settings.

Spatial planning is a highly political process in which data and information on the suitability or priority of land for conservation or development is only part of the process. Trade-offs between conservation and development of land involve trade-offs between beneficiaries (for example employees of a mine *vs.* those whose water is affected by its operation), trade-offs over time (increased crop productivity from the implementation of intensive agriculture now *vs.* long term soil exhaustion or salinisation) and trade-offs over space (upstream beneficiaries of a reforestation scheme *vs.* those downstream whose annual water flows might decline as a result). Trade-offs can be manifest in economic terms, in livelihood opportunities, in health and welfare and in environmental terms. Managing these kinds of tradeoffs requires good spatial information on conservation and development priority that is made available in a usable form for application locally and regionally and in a consistent manner between regions, globally as has been the focus here and more generally in the approach adopted by Co$ting Nature.

Acknowledgements

Co$ting Nature is a collaborative project of King's College London, UNEP-WCMC and AmbioTEK Community Interest Company. All data providers are gratefully acknowledged.

References

Bianchi, F.J.J.A., Booij, C.J.H. and Tscharntke, T. 2006. Sustainable pest regulation in agricultural landscapes: a review on landscape composition, biodiversity and natural pest control. Proceedings of the Royal Society B: Biological Sciences 273(1595): 1715–1727.

Birdlife. 2012. Birdlife International IUCN redlist for birds. http://www.iucnredlist.org/technical-documents/spatial-data#birds.

Bruijnzeel, L.A., Mulligan, M. and Scatena, F.S. 2011. Hydrometeorology of tropical montane cloud forests: Emerging patterns. Hydrological Processes 25(3): 465–498.

Center for International Earth Science Information Network (CIESIN). 2002. Country-level GDP and Downscaled Projections based on the A1 A2 B1 and B2 Marker Scenarios 1990–2100 [digital version]. Palisades NY: CIESIN Columbia University. Available at http://www.ciesin.columbia.edu/datasets/downscaled (date accessed).

Costanza, Robert D'Arge, Ralph Groot, Rudolf de Farber, Stephen Grasso, Monica Hannon, Bruce Limburg, Karin Naeem, Shahid O'Neill, Robert V. Paruelo, Jose Raskin, Robert G. Sutton, Paul Belt and Marjan van den. 1997. The value of the world's ecosystem services and natural capital. Nature 387(6630): 253–260.

Dapples, F., Lotter, A.F., van Leeuwen, J.F., van der Knaap, W.O., Dimitriadis, S. and Oswald, D. 2002. Paleolimnological evidence for increased landslide activity due to forest clearing and land-use since 3600 cal BP in the western Swiss Alps. J. Paleolimnol. 27(2): 239–248.

De Groot, Rudolf S., Matthew A. Wilson and Roelof M.J. Boumans. 2002. A typology for the classification, description and valuation of ecosystem functions, goods and services. Ecological Economics 41(3): 393–408.

Dilley, M. 2005. Natural Disaster Hotspots: A Global Risk Analysis. Version 1.0. Disaster Risk Management Series, No 5. World Bank, Washington DC.

FAO. 2010. GIEWS: World road trails. Whole world's roads and railways [http://ldvapp07.fao.org:8030/downloads/layers/world_roadstrail.xml].

FAO. 2012. Agro-ecological suitability and productivity. Crop suitability index (class).

Ghilarov, Alexei, M. 2000. Ecosystem functioning and intrinsic value of biodiversity. Oikos 90(2): 408–412.

Hansen, M., DeFries, R., Townshend, J.R., Carroll, M., Dimiceli, C. and Sohlberg, R. 2006. Vegetation Continuous Fields MOD44B, 2001 Percent Tree Cover, Collection 4, University of Maryland, College Park, Maryland, 2001.

Ho, L.C. 1976. Variation in the carbon/dry matter ratio in plant material. Ann. Bot. 40(1): 163–165.

IUCN, Conservation International, and NatureServe. 2008a. An analysis of amphibians on the 2008 IUCN Red List [www.iucnredlist.org/amphibians]. Downloaded on 9 May 2009.

IUCN, Conservation International, Arizona State University, Texas A&M University, University of Rome, University of Virginia, and Zoological Society London. 2008b. An analysis of mammals on the 2008 IUCN Red List [www.iucnredlist.org/mammals]. Downloaded on 9 May 2009.

IUCN. 2010. An analysis of reptiles on the 2010 IUCN Red List [http://www.iucnredlist.org/technical-documents/spatial-data]. Downloaded on 29 November 2010.

Justus, James, Colyvan, Mark, Regan, Helen and Maguire, Lynn. 2009. Buying into conservation: intrinsic versus instrumental value. Trends Ecol. Evol. 24(4): 187–191.

Kier, G. and Barthlott, W. 2001. Measuring and mapping endemism and species richness: a new methodological approach and its application on the flora of Africa. Biodiversity and Conservation 10: 1513–1529.

LandScanTM Global Population Database. 2007. Oak Ridge, TN: Oak Ridge National Laboratory. Available at http://www.ornl.gov/landscan/.

Lehner, B. and Doll, P. 2004. Development and validation of a global database of lakes, reservoirs and wetlands. J. Hydrol. 296(1-4): 1–22.

Lehner, B., Verdin, K. and Jarvis, A. 2008. New global hydrography derived from spaceborne elevation data. Eos, Transactions, AGU 89(10): 93–94.

Mulligan, M. 2013. Co$ting Nature Policy Support System. Available at www.policysupport.org/costingnature.

Mulligan, M.A., Guerry, Arkema, K., Bagstad, K. and Villa, F. 2010. Capturing and quantifying the flow of ecosystem services. pp. 26–33. *In*: Silvestri, S. and Kershaw, F. (eds.). Framing the Flow: Innovative Approaches to Understand, Protect and Value Ecosystem Services Across Linked Habitats. UNEP World Conservation Monitoring Centre, Cambridge, UK [available online http://www.unep.org/pdf/Framing_the_Flow_lowres_20final.pdf].

Mulligan, M. 2009. The human water quality footprint: agricultural, industrial, and urban impacts on the quality of available water globally and in the Andean region. *In*: Proceedings of the International Conference on Integrated Water Resource Management and Climate Change, Cali, Colombia. 11 pp.

Mulligan, M. 2009b. Global mean dry matter productivity based on SPOT-VGT (1998–2008). http://geodata.policysupport.org/dmp.

Mulligan, M., Saenz, L. and van Soesbergen, A. 2011. Development and validation of a georeferenced global database of dams. Submitted Water Resources Research. http://www.ambiotek.com/dams.

Mulligan, M. 2010. SimTerra: A consistent global gridded database of environmental properties for spatial modelling. http://www.policysupport.org/simterra.

Mulligan, M. 2010a. A combined global database of mines. http://www.ambiotek.com/mines.

Mulligan, M. 2010b. A combined global database of oil and gas wells. http://www.ambiotek.com/oilandgas.

National Geophysical Data Center/World Data Center (NGDC/WDC) (2011) Historical Tsunami Database, Boulder, CO, USA (Available at: http://www.ngdc.noaa.gov/hazard/tsu_db.shtml).

Oksanen, Markku. 1997. The moral value of biodiversity. Ambio 26(8): 541–545.

Peña-Arancibia, J.L., van Dijk, A.I., Guerschman, J.P., Mulligan, M., Bruijnzeel, L.A. and McVicar, T.R. 2012. Detecting changes in streamflow after partial woodland clearing in two large catchments in the seasonal tropics. J. Hydrol. 416: 60–71.

Purvis, Andy and Hector, Andy. 2000. Getting the measure of biodiversity. Nature 405(6783): 212–219.

Raudsepp-Hearne, C., Peterson, Garry, D. and Bennett, E.M. 2010. Ecosystem service bundles for analyzing tradeoffs in diverse landscapes. Proc. Natl. Acad. Sci. USA. 107(11): 5242–5247.

Ramankutty, N., Evan, A.T., Monfreda, C. and Foley, J.A. 2008. Farming the planet: 1. Geographic distribution of global agricultural lands in the year 2000. Global Biogeochemical Cycles Vol. 22 GB1003 doi:10.1029/2007GB002952.

Ruesch, Aaron and Gibbs, Holly, K. 2008. New IPCC Tier-1 Global Biomass Carbon Map For the Year 2000. Available online from the Carbon Dioxide Information Analysis Center http://cdiac.ornl.gov], Oak Ridge National Laboratory, Oak Ridge, Tennessee.

Saatchi, S., Harris, N.L., Brown, S., Lefsky, M., Mitchard, E.T., Salas, W., Zutta, B.R., Buermann, W., Lewis, S.L., Hagen, S., Petrova, S., White, L., Silman, M. and Morel, A. 2011. Benchmark map of forest carbon stocks in tropical regions across three continents. Proc. Natl. Acad. Sci. U.S.A. 2011 Jun 14; 108(24): 9899-9904.

Scharlemann, J.P.W., Hiederer, R. and Kapos, V. 2009. Global map of terrestrial soil organic carbon stocks. A 1-km dataset derived from the Harmonized World Soil Database. UNEP-WCMC & EU-JRC, Cambridge, UK.

Schneider, A., Friedl, M.A. and Potere, D. 2009. A new map of global urban extent from MODIS data. Environmental Research Letters, volume 4, article 044003.

Siebert, S., Döll, P., Feick, S., Frenken, K. and Hoogeveen, J. 2007. Global map of irrigation areas version 4.0.1. University of Frankfurt (Main), Germany, and FAO, Rome, Italy.

Spalding, M.D., Blasco, F. and Fields, C.D. (eds.). 1997. World Mangrove Atlas. The International Society for Mangrove Ecosystems, Okinawa, Japan. 178 pp.

Tilman, D., Reich, P.B. and Knops, J.M. 2006. Biodiversity and ecosystem stability in a decade-long grassland experiment. Nature 441(7093): 629–632.

Troy, Austin and Wilson, Matthew, A. 2006. Mapping ecosystem services: practical challenges and opportunities in linking GIS and value transfer. Ecological Economics 60(2): 435–449.

USGS. 2006. Shuttle Radar Topography Mission Water Body Dataset, Available online at: http://edc.usgs.gov/products/elevation/swbd.html.

World Database on Protected Areas (WDPA) Annual Release 2012 (web download version), March 2012. The WDPA is a joint product of UNEP and IUCN, prepared by UNEP-WCMC, supported by IUCN WCPA and working with Governments, the Secretariats of MEAs and collaborating NGOs. For further information protectedareas@unep-wcmc.org.

10

Can Investment in River Basins Sustain Global Development of Food and Energy Systems?

Simon Cook[1,*] and *Myles Fisher*[2]

Introduction

The Need for a Considered Analysis of Sustainable Development in River Basins

Hardly a week passes without a warning in the serious press of an impending crisis concerning global food or water resources. The arguments are straightforward: development proceeds to meet the demands of an expanding and more affluent population; land and water resources are finite; history is littered with examples of society's inability to manage finite natural resources. Therefore catastrophe looms.

The arguments seem convincing. Moreover, they are sufficiently pressing to demand attention. Combined with overwhelming evidence of climate change, the potential threat to global food, water and environmental securities demands our attention.

Yet while the overall threat to resources seems unquestionable, the consequences to food or energy security and the sustainability of development seem less certain.

[1] Formerly Director of CGIAR Program Water, Land and Ecosystems based at Colombo, Sri Lanka. Currently President of FFR, Fundacion Futuros Rurales, Cali, Colombia.
Email: simonernest@gmail.com

[2] Emeritus Scientist at CIAT, Centro Internacional de Agricultura Tropical, Cali, Colombia.
Email: mylesjfisher@gmail.com

* Corresponding author

The costs of action to address the problem are large. Uncertainty plays into the hands of the procrastinating politician. Our goal must be to support development that is sustainable and inclusive.

A balanced analysis is required. Rather than merely pointing out the risks of possible catastrophe (headline-grabbing as they surely are), we try to explain the interplay between development processes and the ecosystem resources from river basins on which these processes depend. We believe this is a surer way to understand the problems and to identify the opportunities to avert them.

In this chapter we reviews analyses of the links between development and the river basins that support it. We base this on detailed multi-disciplinary research within ten major river basins in the developing world, which together contain over half the world's poor and represent a wide range of biophysical, economic and social conditions in Latin America, Africa and Asia. We use the inclusive wealth framework of UNEP (2012) to evaluate our observations.

Analysis of Conditions in 10 River Basins

Is growth in river basins sustainable for the foreseeable future? Or can we see evidence that economic development is driving food and energy systems to a point that is likely to result in irreversible and possibly rapid decline of ecosystems?

To assess the likelihood of either outcome we examine the following aspects in ten river basins analysed by the Basin Focal Projects (BFPs) of the CGIAR Challenge Program on Water and Food (CPWF) (Fisher and Cook, 2012a).

- Problems: These often make the headlines, and stir politicians to act. They are important to acknowledge, since they are what stimulates investigation. But they represent events, not the processes that caused them.
- Processes: A description of the processes linking food, water and development systems and their principal outcomes.
- Evaluation: We review characteristics according to the inclusive wealth framework to assess the sustainability of growth.

Problems

Four types of general problems can occur in river basins:

1. Resource scarcity. Population growth, coupled with increases in per capita demand and the economic activity that accompanies it, lead to absolute scarcity of land and water resources. Per capita water availability has reduced in some basins to the point at which it is below the level regarded as critical (Falkenmark, 1997). Land availability in many areas, including India's Punjab, in the Ganges delta, the Yellow River and West Kenya seems insufficient to support farmers' exit from poverty.
2. Resources inaccessibility. Total resource availability remains moderate. However, poor access to resources and unequal distribution constrain opportunities for

many. Poor availability may be caused by lack of infrastructure, or social barriers that block access to resources for specific groups of people.

3. Low resource use productivity. Resources may be available, but users derive little benefit from their use. Users may be locked into low productivity through reasons of lack of co-factors of productivity such as lack of investment capital or poor access to markets.
4. Resource depletion. Inappropriate use may lead to degradation of land, water or biological resources. A decline occurs in total inclusive wealth. This may be irreversible.

Each of these will result in loss of potential livelihood support or sustainability. By and large, they are also the consequences of patterns of development.

Processes: Key Insights from Analysis of River Basins

In this section we summarize the key insights that link development outcomes with attributes and food and water systems. We use examples from the ten BFP river basins.

Many of the BFP basins have high levels of poverty and there are some similarities, but each presents different underlying problems. These problems must be addressed if the goals of increasing food production and overcoming poverty are to be met.

Water scarcity and poverty

Population growth has reduced available water in some basins below 1700 m^3/capita/yr, the level considered secure (Falkenmark, 1989). Absolute water scarcity worsens when the growing population depends on unsustainable irrigation as in the Yellow, Indus, Karkheh and upstream Limpopo basins.

Water can be physically scarce in arid regions or it can be scarce because its quality is poor, whether contaminated by sediment or pollutants from upstream activities, or unclean in the case of untreated urban waste. Water quality is often thought to be an indicator of poverty, but does poor quality water cause poverty or is it a consequence? Poor-quality water causes water-borne diseases and infant mortality. Better quality water will lower infant mortality and in doing so will cause an immediate, voluntary reduction in fertility rates and thus reduce the rate of population increase (Sachs, 2007).

The BFP studies in ten river basins were unable to establish a causal relation between the availability of water per capita and poverty (Kemp-Benedict et al., 2012).

Water quality

For the rural poor in some basins, water quality is more important than quantity. UN data suggests that over a billion people remain without access to safe drinking water. Water quality is a universal issue for the rural poor in the Nile, the Indus-Ganges, and the Volta (Fisher and Cook, 2012b). It is also a problem in the relatively more-developed basins of the Andes, where mining and unsustainable agriculture threaten water quality (Mulligan et al., 2012a). But it is difficult to provide safe water to the widely dispersed populations of the rural poor. Small, ferro-concrete water tanks ("Thai jars") in Nepal provide clean domestic water piped from upstream, with the surplus used to irrigate home gardens. This multiple-use method could be applied more widely. Rainwater harvesting for domestic water receives little attention, but it is viable even in semi-arid countries. The benefits of even modest improvements in water infrastructure seem obvious (Barron et al., 2012), but for the poorest without security of tenure or access to credit, even these may prove unattainable.

Water balance and availability

The status of overall availability of water in a basin, country or region is not a reliable indicator of its availability to individuals. For example commercial farmers in the upper Limpopo basin have preferential access to water resources compared with subsistence farmers in the former "homelands" (Sullivan and Sibanda, 2012).

Analyses of water flows to sub-basin scale within the ten basins by Kirby et al. (2012) and Mulligan et al. (2012b) illustrate the dangers of over-generalization. All basins experience increased demand for water, but the location of increased demand and the interaction of basin hydrology determine the nature of stress that is likely to occur. For example, in the Nile basin Egypt has, until recently, opposed development of water resources by Ethiopia because of acute sensitivity to changes in flow of the Blue Nile. Similarly, the Karkheh basin in Iran is effectively closed, so any further plans for development of water resources have grave consequences for inflow to the Hoor-al-Azim wetlands at its mouth. Conversely Lemoalle et al. (2012) discovered that flow in the Volta is relatively insensitive to development of small reservoirs, on account of increasing contributions of precipitation downstream. The Ganges and Mekong both discharge huge volumes of water (352 and 418 km^3/year respectively) (Mulligan et al., 2012) but nevertheless present problems of sensitivity to development. In the case of the Mekong, it is due to the dependence of so many people on aquatic ecosystems. Their dependence may be compromised by the substantial development of hydro-power, which is ongoing. In the case of the Ganges, the issue of flows past the Farraka Barrage into Bangladesh have been a source of disquiet for decades.

Vulnerability to water-related hazards

Drought, flood and water-borne diseases are hazards that have major impacts on development. They inflict hardship in countries that have little capacity to manage them, the Niger, the Volta and the Nile basins contain Sahelian zones that are frequently subjected to drought. Floods in the Indus were catastrophic with 30% of Pakistan flooded. Flow in the Limpopo is extremely variable and vulnerable to moderate changes in climate (Eastham et al., 2010). It is subject to both drought and floods.

Malaria kills an estimated 6–700,000 people each year, mainly children in Africa (World Health Organization, 2012). This and other water-related diseases remain a major obstacle to development. Prevalence of disease is closely associated with the development of water resources, for example, in Burkina Faso, up to 70% of cases of schistosomiasis is associated with irrigation and small reservoirs (Lemoalle et al., 2012).

Climate change introduces additional uncertainties

Climate change is a threat everywhere, but the poor are more vulnerable as they have few resources to adapt to its effects. Temperatures will rise by 2°C–3°C by 2050, which will increase water losses to evaporation. Higher temperatures will reduce yields of maize and rice, hence water productivity, although by how much is uncertain. It is less certain how precipitation will change, but Mulligan et al. (2012c) anticipate that of the ten basins studied, all except the Limpopo will receive more rainfall though with higher temperatures, the net effect on water balance is less certain. River flows will be more variable with more floods because there will be less snow and ice, which currently even out river flows. Plant breeders may be able to produce crop varieties that are resistant to or tolerant of stresses, but there will probably be yield penalties. There are also agronomic solutions, such as sowing the crop later to avoid the very hot weather that precedes the monsoon and so reduce evaporation and the amount of water the crops need.

Access and Equality

Water scarcity is not simply linked with poverty because of the strong influence of where a basin, country or region is on a development trajectory from agricultural to industrial (World Bank, 2007; Kemp-Benedict et al., 2012). The position on the development trajectory "does not predict the character of water-poverty links, but is such a powerful factor that a first step in analysing the water–food–poverty links within a basin should be to determine where it lies along that trajectory" (Kemp-Benedict et al., 2012). Rural poverty is not eliminated in basins that have achieved industrialized status, but tends to remain as pockets that have not participated in the trend to industrialization. Power relations in a community often determine the access to resources of particular groups (ethnic, gender). Sen (1999) defines poverty as the constraints imposed upon individuals to live the kind of life she or he values.

Productivity and Eco-Efficiency

Water productivity

Cai et al. (2012) review the evidence from river basins and identify that outside some notable hotspots, water productivity, that is, the conversion efficiency with which agricultural systems convert water into food or other benefit, remains well below potential. The hotspots include parts of the Ganges and Yellow River basins, together with the Nile delta. The gap between current and potential water productivity represents a major opportunity for intensification that will be necessary to meet emerging demands for food, without increasing the water consumed by agriculture.

Populations in the sub-Sahel are doubling every 30 years, and seem likely to continue to do so. By increasing the cropped area, food production has kept pace with the increase over the last 20 years, which can continue in the short term. For the longer term, however, the low WP of rainfed agriculture in the sub-Sahel must increase. WP can be increased with appropriate agronomy (high-yielding varieties and fertilizer), but farmers do not use them because of other factors, such as no market access, poor storage and no credit to buy the inputs. "As pressure on the available land increases, however, higher WP is the only solution to providing the food that will be needed with the water that is available" (Fisher and Cook, 2012b).

Eco-efficiency

Ecosystem services (ESs), which Daily (1997) defines as "the conditions and processes by which natural ecosystems, and the species that make them up, sustain and fulfil human life." Building on this definition, we include in ESs all those goods and services that water provides people within a river basin. ESs are not used individually or by individuals, but collectively by communities or organizations.

Eco-efficiency is the amount of agricultural products relative to the ecological resources that are used as inputs. The inputs may be some or all of land, water, nutrients, energy, or biological diversity. Assessment of eco-efficiency should also consider human and economic inputs of capital and labour and should not ignore consequences such as the environmental loads on broader ecosystems. These might include losses of nutrients, salts, acids or sediments to other ecosystems and emissions of greenhouse gases to the atmosphere, or any other negative or positive effect of agriculture on other ecosystem services (Keating et al., 2010).

Understanding the Dynamics of Development

Reviewing the analyses from ten river basins, Cook et al. (2012) suggest that while it is conventional to focus on the problems of resource use that occur as a result of growth, the fundamental requirement to meet current and future demand is for more intense development of land and water resources. Such development needs to be accompanied by a greater awareness of the need to share benefits and risks, and build on a long-term political vision.

"[I]mproved agriculture and water management require technical, sociological, and regulatory changes to address the wider causes of poverty" (Ogilvie et al., 2012). We could say this of all the BFP basins, but how can these changes happen? The answer is economic development, which can take place through support of agriculture: "Agriculture has served as a basis for growth and reduced poverty in many countries, but more countries could benefit if governments and donors were to reverse years of policy neglect and remedy their underinvestment and misinvestment in agriculture" (World Bank, 2007). The *World Development Report* (World Bank, 2007) argues that agriculture was heavily taxed to support industrialization, which reflected political economies dominated by policies that reflect urban interests, which "proved lethal in Africa" (Byerlee et al., 2009).

Agriculture continues to be a fundamental instrument for sustainable development and poverty reduction, even while economies move beyond agriculture and their economies become more industrialized: "The global development agenda will not be possible without explicitly focusing on the role of agriculture for development" (Byerlee et al., 2009).

Smallholder farming needs a productivity revolution if agriculture is to be the basis for economic growth in the agriculture-based countries where water productivity (WP) is low. For agriculture to drive the development agenda, agriculture needs improved governance at local, national and global levels. Growth in GDP from agriculture reduces poverty at least twice as much as growth in GDP in non-agriculture sectors. In China, growth in agriculture reduced poverty 3.5 times more than growth outside agriculture, while in Latin America it was 2.7 times more (World Bank, 2007).

Increased agricultural activity has major impacts on the river basin systems that support it. But as development moves beyond agriculture, demand for water from other sectors increases. Balanced development of water and food systems requires detailed insight of conditions within basins, together with analysis of the processes that drive development (Fisher and Cook, 2012b).

Importance of Effective Institutions

Institutions, which include things like land tenure and social norms as well as organizations from national to municipal governments, are the key to many problems in basins. Acemoglu and Robinson (2012) assign many of the failures of development to institutional problems. There are transnational considerations that affect whole countries or provinces, and there are institutions that operate at national level down to individual towns or villages.

Transnational institutions

Weak trans boundary institutions are common, identified in Fisher and Cook (2012a) in all but the Karkheh Basin, which is entirely within Iran. But even where a basin is in one country, there can be problems, as in China and India where provinces and states defend their rights to the waters within their borders. Nevertheless, Giordano et al. (2005) show that despite tensions, trans boundary rivers are more a subject of agreement than conflict. The papers in Fisher and Cook (2012a) support that conclusion.

Salman (2010) concludes that co-operation amongst riparian countries is the cardinal principle of the law of international waters, and that the interests and concerns of both upstream and downstream riparian countries need to be considered by all parties. Implementing that conclusion needs good will and transparency, which is unlikely if countries pursue their own agendas, as they usually do.

Most "transnational rivers do have a statutory institution, nominally with a coordinating role, but the participating countries in general have not ceded any useful authority to the institutions they have created. They remain bodies that support dissemination of research, and convene conferences and meetings, but they do little to influence political outcomes, which can only be arrived at by consensus of the constituent countries" (Fisher and Cook, 2012b).

National institutions

Fisher et al. (2012) identified strengths and weaknesses of institutions to generalise why they fail to address basin-wide issues of water, poverty, and livelihoods. They showed how a more comprehensive, integrated approach might change them to be more broadly relevant to basin-wide needs. Institutions might then address the mismatch between development and the need to provide ecosystem services relevant to food, poverty, livelihoods and sustainable ecosystems.

Institutions and organizations are key to solving the problems of water and food caused by the increasing global population. But the analysis of the nine basins (excluding the São Francisco) reviewed by Fisher and Cook (2012a) shows that the problem is not resource constraint but use of resources and distribution of the rights and benefits. "It is institutions and organizations that determine the benefit derived from use of [ecosystem services], the way the benefits and the resources are shared, and the penalties for abuse" (Fisher and Cook, 2012b).

Institutions and organizations are successful when they meet the demands of the populations they service. But they rarely do so, even though they continue to evolve. Where institutions are divided into separate jurisdictions, "conflict over food, water and livelihoods; . . . conflicts between local interests and the benefit of the whole (including the supporting ecology), and power inequalities" will occur (Fisher and Cook, 2012b).

Evaluation: Assessing Development Against the Inclusive Wealth Framework

The inclusive wealth of river basin systems

UNEP (2012) propose the concept of inclusive wealth to represent the full costs and benefits of economic development. They point out that systems of assessing development that focus only on economic measures such as GDP fail to take account of social and natural capitals on which such wealth depends. They show that bankrupting the natural environment for economic gain may leave a country worse off than before development occurred.

The concept focuses on the assessment of three forms of capital: natural, manufactured and human. Natural capital comprises the stock from which ecosystem good and services are realized. Costanza et al. (1998) estimated the value of these goods and services at the equivalent of over US$ 30 trillion annually. River basins occupy virtually all the terrestrial landscape and so may be considered the source for all ecosystem services. But some, such as water flow, are dependent on the function of river basins as hydrological entities and so we focus on these. These include *supporting* ecosystem services of nutrient dispersal; *provisioning* ecosystem services of water supply for irrigation, rainfed agriculture, industrial or domestic consumption; *regulating* services such as climate regulation or water purification; and cultural services, which comprise a huge diversity of non-material benefits from river basins on which humanity depends.

Manufactured capital includes all material assets generated directly and indirectly by human endeavour. These include roads, buildings and infrastructure from which food and fibre is produced using green and blue water resources. Hydropower assets would also be included.

Human capital comprises the wealth of organizations and institutions that mediate the flow of investment of labour and finance and that control human activity. Also included are knowledge and health assets that increase a person's productivity. At a broad level, they describe the ability of societies to capture, develop and share natural resources and the manufactured wealth they produce. Human capital in its more rustic form appears as the provision of basic health, education and administrative and protective services (World Bank, 2007). The provision of such services is recognized as a pre-requisite for early stages of development. As economies develop, theorists surmise that most gains derive from intellectual products derived from human capitals.

The concept of inclusive wealth offers a major advantage over alternative theoretical frameworks that consider only economic or livelihood capitals; namely that it accepts that some loss of natural capital is inevitable during development, but that the aim of development is to ensure that this loss is converted to equal or greater gain of manufactured and human capitals. In this way, decision makers are provided with the flexibility to decide how best to use natural capital [which is inevitable], which natural capital should be preserved and how well economic development has delivered overall benefits from river basins.

Valuation remains a contentious issue, but we believe the approach offers a useful conceptual basis on which to perceive 'good' or 'bad' development. 'Good' development of a river basin's ecosystem services is that which values all capitals and identifies stocks and transfers of capitals in informed and transparent debate. 'Bad' development is that in which certain groups are excluded– that is undervalued—in which the benefits of manufactured capital such as hydropower or intensive agricultural production are focussed on a few at the cost of the natural capital it consumes. 'Good' development identifies the likely long-term outcomes of change. 'Bad' development tends to respond to shorter-term cycles of personal enrichment or politics. In short, the inclusive wealth approach accepts that decision about the use of ecosystem resources in river basins are made according to a range of influences. These cannot be controlled, but they should at least be evaluated.

Applying the Concept to Development in River Basins

Biophysical endowments provide the natural capital on which development in river basins depends, especially during its early phases. For example, analysis from the Niger basin showed how problems in gaining secure access to resources obstruct development. Poor access to water or land resources in other basins hinders farmers' investment in food systems. The health risks that accompany lack of access to safe water remain a widespread hazard throughout the developing world, even while levels of provision and sanitation have improved.

In all cases, problems occur when population growth and development of certain sectors of the economy—especially resource-extractive activities such as extensive cropping, irrigation, forestry or mining—reduce the availability of natural resources to others.

According to the concept of the environmental Kuznets curve (EKC), loss of natural capital increases, and then decreases in intensity as economies become wealthier. The hypothesis expects there to be little initial destruction of natural capital when population pressures are low. Destruction intensifies during intermediate stages of development if the power to consume precedes the development of institutional and political control. Finally, the hypothesis predicts that political and societal pressures enhance the protection of natural resources, though not before there is substantial loss.

Empirical evidence to support the EKC seems elusive although, to be fair, the data sought does not conclusively disprove the idea. Stern (2004) dismisses the class EKC, but points out that developing countries may be short-circuiting the curve by adopting the environmental protection standards normally expected of developed countries. Even without proper valuation, there seems overwhelming evidence of the loss of natural capital. Large areas have degraded land and water resources and in sub-Sahara Africa chronic underinvestment in fertilizer has resulted in a substantial decline in soil fertility. What is less clear is evidence of reversal or investment in natural capital during later stages of development, although destruction does slow as development proceeds.

Societies consume natural capital to generate manufactured capital, which includes agricultural production. The effectiveness with which this occurs is highly variable. At one extreme we see resource scarcity in the Yellow River basin pushing land and water productivity to levels that seem difficult to improve. In some areas, production has shifted to higher value products. On the other extreme, levels of water productivity over much of sub-Saharan Africa remain extremely low. Increases in food production have come from extension of the cultivated area, not from increases in productivity. Total biomass production is insufficient over large areas to support conservation farming.

The third class of investment is human or social capital. The growth of social capital is reflected in indices of human development, which aggregate measures of poverty, health, education and infrastructure.

We illustrate the concepts with examples from the ten river basins (Table 10.1).

Likely Future Scenarios

Is future growth likely to be sustainable? For some (Rockstrom et al., 2009) systems are approaching irreversible tipping points, after which decline is inevitable. For others (Pingali, 2012) the future seems more positive, and potentials are limited mainly by manageable constraints. Both scenarios could occur, depending on the ability of governments at local, national and international scales to direct investment to increase inclusive wealth, and to ensure that natural capitals—a nation's heritage —is safeguarded to protect vital ecosystem function (Table 10.2).

Conclusions: Is Inclusive Wealth Increasing in River Basins?

In the preceding sections of this chapter we explore analysis from ten river basins. We now assemble this evidence to try to conclude if and where river basins see investment in inclusive wealth. That is, where we see evidence for balanced development that not only increases manufactured and human capitals, but also maintains the natural capital that will make such development sustainable.

Table 10.1. Summary of effects on capitals in basins.

	Basin	Capitals		
		Natural	**Manufactured**	**Human**
Industrial	Andes	Loss of biodiversity. Mineral extraction boom fuelling economy. Low population pressure	Agriculture no longer a major contributor to GDP growth	Widespread inequality
				Political violence
				Social organization required for PES/BSM
	São Francisco	Conflict over dry middle. Consequences for Sobradinho dam	Massive expansion of productivity from Cerrados to agricultural revolution. Commercial agriculture increasing contribution to GDP	EMBRAPA and generous finance supported the agricultural revolution
		Forest loss. Soil degradation overcome		Efforts to overcome inequality
Transitional	Karkheh	Widespread land degradation	Rural economy strongly subsidised	Politics dominates everything
		Water scarcity exacerbated by plans for hydropower		
	IGB	Ganges very different from Indus	Western Ganges very high production, Eastern Ganges, very low productivity	Very strong rural politics. Political hi-jacking obstructs change
		Ganges hydrologic 'power house', extremely high flows	Indus lower productivity than Ganges	Major reforms envisaged in India. Pakistan and Bangladesh success despite poor governments
		Indus highly vulnerable to flow dependency. Over use of ground water	Agriculture uncertain value to GDP growth though huge requirement for food security	
	Mekong	Major loss of aquatic system function	Massive plans for hydropower as part of policies for green energy. Major growth of aquaculture in place of capture fisheries	Rapid expansion but opaque politics leaves many behind
Transitional	Yellow	Extreme water scarcity. Flow ceased in 1998. Degradation on loess plateau controlled	Very high productivity. Now changing	Exceptional institutional strength enables rapid change. Water trading being explored
			Non agriculture rural economy key to success	

Table 10.1. contd....

Table 10.1. contd.

	Basin	Capitals		
		Natural	Manufactured	Human
Agricultural	Limpopo	Land degradation. Water scarcity. Floods and droughts	Mining trumps agriculture. Low productivity outside commercial agriculture areas	Major political changes in place, but colonial legacy difficult to overcome
	Niger	Soil fertility decline. Groundwater resources unknown. Seasonal river flow. Rapid population growth	Drought risk a major impediment to investment	Some of the world's poorest nations. Need to get the basics right [health and education]. Migration a major factor
	Nile	Soil fertility decline	Rapid growth of Ethiopia creates strains; Egypt major food importer; Kenya and Uganda rapid gains post 2000	HDI low but climbing rapidly. Very high population growth exerts severe strains on growth. Indicative
		Egypt and North Sudan totally dependent on upstream supply		
	Volta	Soil fertility decline	Drought obstructs productivity. Currently very low	Major problems caused by shift through agricultural phases
		Stress upstream	Agriculture still dominates in rural areas	Migration a factor
		Groundwater resources available		

Table 10.2. Future conditions will demonstrate characteristics of the 'good' and 'bad' scenarios.

Good	Bad
Parsimonious use of natural capital, agreed with others affected by change and according to agreed values	Appropriation for exclusive use. Excessive or uncontrolled exploitation with no reference to loss of value
Efficient conversion into manufactured capital [e.g., high water productivity]	Low conversion efficiency
Transferrable gains, increase in adaptive capacity	Change 'locks in' people to a situation from which it may be difficult to escape
Change anticipates crisis	Change occurs after the crisis has hit
High levels of organization to share benefits and risks	Poor learning. Organization after the change no better than before

The urbanized basins are the Andean basins, the São Francisco and the Karkheh. Parts of the Limpopo may also be considered urbanized, as is the Nile delta region (see Fig. 10.1). We have difficulty to draw overall conclusions about the development of inclusive wealth in this group of basins.

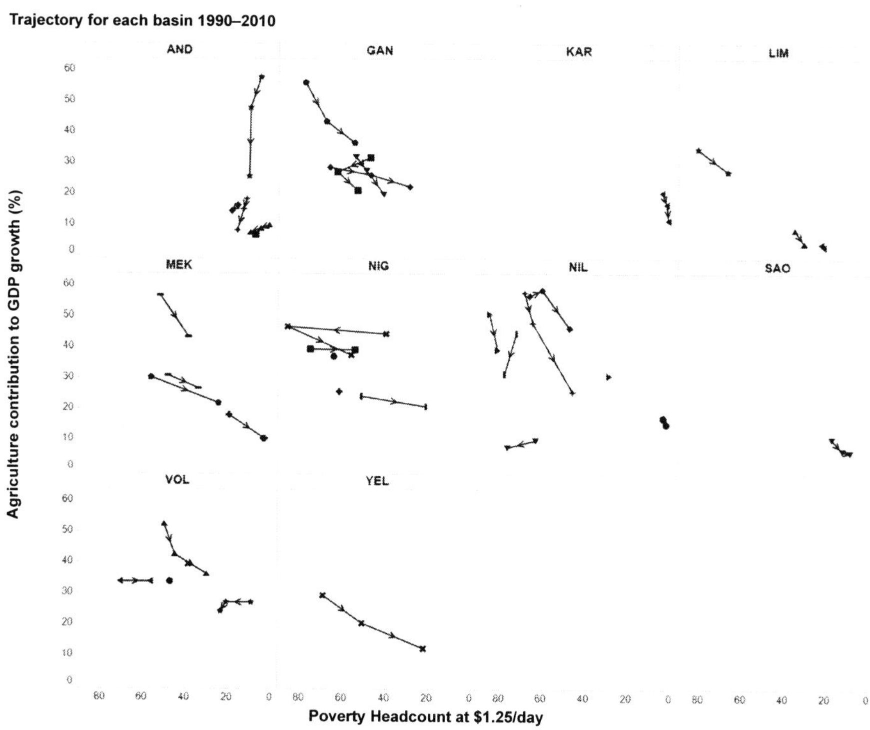

Figure 10.1. Development trajectories for 10 basins. Curves are based on national data for each basin from the World Bank. Countries include: Andes (AND; Bolivia, Colombia, Ecuador, Peru); Indus-Ganges (GAN, Bangladesh, Nepal); Karkheh (KAR), Iran; Limpopo (LIM; Botswana, Mozambique and South Africa); Mekong (MEK; Cambodia, Lao, Vietnam, Thailand); Niger (NIG; Mali, Niger, Cameroun, Nigeria); Nile (NIL; Burundi, Rwanda, Kenya, Egypt); São Francisco (Brazil); Volta (Burkina Faso, Ghana, Benin); Yellow River (YEL, China).

The basins in Latin America are atypical since population pressure and population growth are both low. Social and economic inequalities hinder the acquisition of human capital, but there is substantial evidence of support for health, education and other social services. Mountainous terrain obstructs the development of infrastructure here, but investment in infrastructure in Brazil is also less than 2% of GDP, less than the level required. Nevertheless, development in Brazil has proceeded well, whereas that in the Andean basins appears to have faltered (Fig. 10.1).

A strong growth of mining in Andean basins countries threatens natural capital. The threat is compounded by climate change that is expected to change cropping patterns, though this is partially countered by national policies for adaptation. Part of the Cerrados of Brazil, which has seen major expansion of agricultural productivity (manufactured capital), lie in the São Francisco basin. Natural capital in the Karkheh basin in Iran is seriously damaged by over grazing and soil erosion. This basin is virtually closed. There are parallels between the concerns over flows in the Karkheh to those in the São Francisco, which has seen strong political protest in recent years over plans to extend irrigation to dry areas of NE Brazil.

Conditions within these basins vary widely. Natural capital in the Andean basins is reducing; those in the São Francisco and Karkheh are already low. Agricultural and non-agricultural productivity is higher in Latin America than in the Karkheh. All basins have active policies to support the rural poor and retrieve human capital from conditions of political uncertainty. Brazil has also succeeded in building a strong economy to develop manufactured capital. In other basins, this basis for growth appears weaker.

Basins in transitional economies include the Yellow River, the IGB and the Mekong. The first two have supported intensive development for decades, which has exhausted natural capital in some cases. This includes the well-documented incidence in 1998 when the Yellow River ceased to reach the sea for over 200 days. The Indus also is virtually closed. The Ganges has a huge average annual flow of 350 km^3/year and yet intensive and uncontrolled consumption in the western part of the basin is causing problems with deficit use of groundwater. Plans to develop water resources in the eastern Ganges seem likely to raise concerns to downstream Bangladesh.

These areas contrast strongly with the Mekong, which until recently was undeveloped. All this has changed with strong growth of the regional economy and with it seemingly insatiable demands for hydropower. This change threatens the natural capital of the aquatic ecosystems on which many depend, especially the poorest.

Has the consumption of natural resources led to an equal or greater production of manufactured and human capital? In China it has, because of the massive gains in productivity and human livelihoods (UNEP, 2012). This is supported by the spectacular decline in poverty in the last 20 years (Fig. 10.1). Agricultural water productivity is also very high over large areas of the Yellow River basin. Population pressure seems likely to ease as a consequence of the one child policy. Policy is strong and leads to rapid change. But is this more or less resilient?

While parallels are often drawn between China and India, the situation in the IGB differs in key attributes. First the productivity gained from natural capital is lower. Second, exploitive use of natural capital such as massive overuse of groundwater in the Indus and western Ganges is politically difficult to control. Third, some authors doubt that manufactured capital can continue to grow without major reforms (Economist,

2012). Finally, can India maintain investment in the face of continued population growth?

Despite these concerns, we are cautiously optimistic. These basins have made major gains in wellbeing, albeit with substantial irreversible cost to natural capital. There are two questions: Does sufficient natural capital remain to prevent systems from passing their 'tipping point', at which they are liable to catastrophic collapse. The other is whether sufficient human capital is being created to ensure continued, but less consumptive growth.

The final class is those basins with agricultural economies. These include most of the Limpopo, the Niger, most of the Nile and the Volta basins. Additionally, Nepal in the Ganges and the Lao PDR in the Mekong would be classified as agricultural. The patterns of development diverges very strongly from the 'winners', such as Ghana, Cameron, Uganda and Ethiopia to 'losers' such as Mali, Burundi and Cote d'Ivoire.

Natural capital in these basins is mostly unexploited. Water resources, including groundwater resources, are undeveloped, agricultural land receives very low inputs, resulting in a net decline of soil fertility. Conversely, a boom in mineral resources fuels growth in some areas. Agricultural productivity is extremely low in all of these basins, with static crop yields over the past 40 years.

Stagnating yields are a major bottleneck to rural development but are potentially also the greatest opportunity. Because the current yields are so low, it is technically possible for very large relative increases of production over very large areas. What is required is that enabling conditions be put in place. Were this to occur, it would influence global sustainability because a production hike from rainfed systems would generate more manufactured capital without decreasing natural capitals.

We are left with development of human capital, at present extremely low in basins such as the Niger. Reports paint a picture of African economies emerging from a period of political instability with high growth rates (Roxburgh et al., 2010) but with population also growing rapidly. Moreover, the improvements are patchy (Fig. 10.1). Not all economies in these basins have reached a point of stability that can foretell continued growth. Continued investment in human capital could be enabled by rapid growth of manufactured capitals, if systems can maintain political stability and keep ahead of the pressures from population growth and climate change.

References

Acemoglu, D. and Robinson, J.A. 2012. Why Nations Fail: The Origins of Power, Prosperity and Poverty. Crown Business, New York.

Byerlee, D., de Janvry, A. and Sadoulet, E. 2009. Agriculture for development: toward a new paradigm. Ann. Rev. Res. Econ. 1: 15–31.

Cai, Xueliang, Molden, David, Mainuddin, M., Sharma, Bharat, Ahmad, M.D. and Karimi, Poolad. 2012. Producing more food with less water in a changing world: assessment of water productivity in 10 major river basins. pp. 42–62. *In*: Fisher, M. and Cook, Simon (eds.). Water, Food and Poverty in River Basins: Defining the Limits. Routledge, London.

Costanza, R., dArge, R., de Groot, R., Farber, S., Grasso, M., Hannon, B., Limburg, K., Naeem, S., Oneill, R.V., Paruelo, J., Raskin, R.G., Sutton, P. and van den Belt, M., 1997. The value of the world's ecosystem services and natural capital. Nature 387: 253–260.

Daily, G.C. 1997. What are ecosystem services? pp. 1–10. *In*: Dailey, G.C. (ed.). Nature's Services: Societal Dependence on Natural Ecosystems. Island Press, Washington DC.

Eastham, J., Kirby, M., Mainuddin, M. and Thomas, M. 2010. Water use accounts in CPWF basins: Simple water use accounting of the Limpopo Basin. CPWF Working Paper Basin Focal Project Series, BFP06. Colombo, Sri Lanka: The CGIAR Challenge Program for Water and Food. 21 pp.

Economist. 2012. Indian reform: At last. The Economist, London. Sep. 22nd 2012.

Falkenmark, M. 1989. The massive water scarcity now threatening Africa: why isn't it being addressed? Ambio 18: 112–118.

Falkenmark, M. 1997. Society's interaction with the water cycle: a conceptual framework for a more holistic approach. Hydrol. Sci. J. 42: 451–466.

Fisher, M. and Cook, S. 2012a. Water, Food and Poverty in River Basins: Defining the Limits. Routledge, London.

Fisher, M. and Cook, S. 2012b. Introduction. pp. 1–10. *In*: Fisher, M. and Cook, S. (eds.). Water, Food and Poverty in River Basins: Defining the Limits. Routledge, London.

Fisher, M.J., Cook, S., Tiemann, T. and Nickum, J.E. 2012. Institutions and organizations: the key to sustainable management of resources in river basins. pp. 846–860. *In*: Fisher, M. and Cook, S. (eds.). Water, Food and Poverty in River Basins: Defining the Limits. Routledge, London.

Giordano, M.F., Giordano, M.A. and Wolf, A.T. 2005. International resource conflict and mitigation. J. Peace Res. 1: 47–65.

Keating, B.A., Carberry, P.S., Bindraban, P.S., Asseng, S., Meinke, H. and Dixon, J. 2010. Eco-efficient agriculture: concepts, challenges, and opportunities. Crop Science 50: 109–119.

Kemp-Benedict, E., Cook, S., Allen, S.L., Vosti, S., Lemolle, J., Giordano, M., Ward, J. and Kaczan, D. 2012. Connections between poverty, water, and agriculture: evidence from ten river basins. pp. 1–18. *In*: Fisher, M. and Cook, S. (eds.). Water, Food and Poverty in River Basins: Defining the Limits. Routledge, London.

Kirby, M., Mainuddin, M. and Eastham, J. 2010. Water-use accounts in CPWF basins: Model concepts and description. CPWF Working Paper: Basin Focal Project series, BFP01. Colombo, Sri Lanka: The CGIAR Challenge Program on Water and Food. URL: http://cgspace.cgiar.org/handle/10568/4084.

Lemoalle, J. and de Condappa, D. 2012. Farming systems and food production in the Volta Basin. Water International 35: 192–217.

Mulligan, M., Rubiano, J., Hyman, G., White, D., Garcia, J., Saravia, M., Leon, J.G., Selvaraj, J.J., Gutierrez, T. and Saenz-Cruz, L.L. 2012a. The Andes 'basin': biophysical and developmental diversity in a climate of change. pp. 472–492. *In*: Fisher, M. and Cook, S. (eds.). Water, Food and Poverty in River Basins: Defining the Limits. Routledge, London.

Mulligan, M., Saenz Cruz, L.L., Pena-Arancibia, J. and Fisher, M. 2012a. Water availability and use across basins. pp. 17–41. *In*: Fisher, M. and Cook, S. (eds.). Water, Food and Poverty in River Basins: Defining the Limits. Routledge, London.

Mulligan, M., Fisher, M., Sharma, B., Xu, Z.X., Ringler, C., Mahé, G., Jarvis, A., Ramírez, J., Clanet, J.-C., Ogilvie, A. and Mobin-ud-Din, Ahmad. 2012. The nature and impact of climate change in the Challenge Program on Water and Food (CPWF) basins. pp. 96–124. *In*: Fisher, M. and Cook, S. (eds.). Water, Food and Poverty in River Basins: Defining the Limits. Routledge, London.

Ogilvie, A., Mahé, G., Ward, J., Serpantié, G., Lemoalle, J., Morand, P., Barbier, B., Tamsir Diop, A., Caron, A., Namarra, R., Kaczan, D., Lukasiewicz, A., Paturel, J.E., Liénou, G. and Clanet, J.C. 2012. Water agriculture and poverty in the Niger River basin. pp. 594–622. *In*: Fisher, M. and Cook, S. (eds.). Water, Food and Poverty in River Basins: Defining the Limits. Routledge, London.

Pingali, P. 2012. Green Revolution: Impacts, limits, and the path ahead. PNAS 109: 12302–12308.

Rockström, J., Steffen, W., Noone, K., Persson, Å., Chapin, III, F.S., Lambin, E., Lenton, T.M., Scheffer, M., Folke, C., Schellnhuber, H., Nykvist, B., De Wit, C.A., Hughes, T., van der Leeuw, S., Rodhe, H., Sörlin, S., Snyder, P.K., Costanza, R., Svedin, U., Falkenmark, M., Karlberg, L., Corell, R.W., Fabry, V.J., Hansen, J., Walker, B., Liverman, D., Richardson, K., Crutzen, P. and Foley, J. 2009. Planetary boundaries: exploring the safe operating space for humanity. Ecology and Society 32: URL: http://www.ecologyandsociety.org/vol14/iss2/art32/.

Sachs, J. 2007. Bursting at the seams: the BBC Reith lectures, 2007. London: Radio 4, British Broadcasting Corporation. Available from: http://www.bbc.co.uk/radio4/reith2007/ [Accessed 22 August 2010].

Salman, S.M.A. 2010. Upstream riparians can also be harmed by downstream riparians: the concept of foreclosure of future uses. Water International 35: 350–364.

Sen, A. 1999. Development as Freedom. Anchor Books, New York.

Stern, D. 2004. The rise and fall of the environmental Kuznets curve. World Development 32: 1419–1439.

Sullivan, A. and Sibanda, M.L. 2010. Vulnerable populations, unreliable water and low water productivity: a role for institutions in the Limpopo Basin. pp. xx-xx. *In*: Fisher, M. and Cook, S. (eds.). Water, Food and Poverty in River Basins: Defining the Limits. Routledge, London.

UNEP. 2012. Inclusive Wealth Report 2012. Measuring Progress toward Sustainability. Cambridge University Press, Cambridge.

World Bank. 2007. World development report 2008: Agriculture for Development. World Bank, Washington, DC.

World Health Organization. 2012. World malaria report: 2012. Geneva: World Health Organization. Available from: http://www.who.int/malaria/publications/world_malaria_report_2012/wmr2012_no_profiles.pdf.

Index